I0040703

Biocatastrophe Lexicon

An Epigrammatic Journey through the Tragedy of our Round-World Commons

Ephraim Tinkham

Engine Company No. 9

Radscan-Chemfall

Est. 1970

Phenomenology of Biocatastrophe

Publication Series Volume 2

ISBN 10: 0-9769153-9-1
ISBN 13: 978-0-9769153-9-3
Davistown Museum © 2010, 2013

Third edition, second printing

Cover photo by the Associated Press

Engine Company No. 9
Radscan-Chemfall
Est. 1970

Disclaimer

Engine Company No. 9 relocated to Maine in 1970. The staff members of Engine Company No. 9 are not members of, affiliated with, or in contact with, any municipal or community fire department in the State of Maine.

This publication is sponsored by

Davistown Museum

Department of Environmental History

Special Publication 69

www.davistownmuseum.org

Pennywheel Press
P.O. Box 144
Hulls Cove, ME 04644

Other publications by Engine Company No. 9

Radscan: Information Sampler on Long-Lived Radionuclides

A Review of Radiological Surveillance Reports of Waste Effluents in Marine Pathways at the Maine Yankee Atomic Power Company at Wiscasset, Maine--- 1970-1984: An Annotated Bibliography

Legacy for Our Children: The Unfunded Costs of Decommissioning the Maine Yankee Atomic Power Station: The Failure to Fund Nuclear Waste Storage and Disposal at the Maine Yankee Atomic Power Station: A Commentary on Violations of the 1982 Nuclear Waste Policy Act and the General Requirements of the Nuclear Regulatory Commission for Decommissioning Nuclear Facilities

Patterns of Noncompliance: The Nuclear Regulatory Commission and the Maine Yankee Atomic Power Company: Generic and Site-Specific Deficiencies in Radiological Surveillance Programs

RADNET: Nuclear Information on the Internet: General Introduction; Definitions and Conversion Factors; Biologically Significant Radionuclides; Radiation Protection Guidelines

RADNET: Anthropogenic Radioactivity: Plume Pulse Pathways, Baseline Data and Dietary Intake

RADNET: Anthropogenic Radioactivity: Chernobyl Fallout Data: 1986 – 2001

RADNET: Anthropogenic Radioactivity: Major Plume Source Points

Integrated Data Base for 1992: U.S. Spent Fuel and Radioactive Waste Inventories, Projections, and Characteristics: Reprinted from October 1992 Oak Ridge National Laboratory Report DOE/RW-0006, Rev 8

An excerpt from *Essays on Biocatastrophe: Phenomenology of Biocatastrophe Publication Series Volume 1*:

"The feed stocks for most endocrine disrupting chemicals are derived from the production of coal, oil, and natural gas… It is clear that endocrine disruption, like climate change, is a spin-off of society's addiction to fossil fuels. Setting aside the effects of endocrine disruptors on infertility, and just considering their influence on intelligence and behavior alone, it is possible that *hormone disruption could pose a more imminent threat to humankind than climate change.*" (TEDX 2009, http://www.endocrinedisruption.com/endocrine.fossilfuel.php).

Biocatastrophe
Lexicon

An Epigrammatic Journey
Through the Tragedy of our Round-World Commons

Volume 2 Table of Contents

Preface

This is the second volume in the three volume *Phenomenology of Biocatastrophe* publication series. This publication series explores the biohistory of the imposition of human ecosystems (industrial, social, political, and economic) on or within natural ecosystems. This series is, in essence, the narration of selected stories about what humans are doing to the biosphere. The essays, definitions, databases, editorial opinions, etc. in these texts explore the impact of human activities on the viability of natural ecosystems in our vulnerable, finite, World Commons. No environmental issue is more important than the unfolding tragedy of the depletion of potable water supplies and the contamination of the atmospheric water cycle. The concomitant spectacle of developed, and now aging, western market economies in crisis is intimately connected with the evolving tragedy of the mad rush to oblivion of an out-of-control global consumer society in a biosphere with finite natural resources.

This is the hard copy edition of our frequently updated online series. Suggestions, corrections, and additional information are welcomed. Please send your feedback to the editor at:

CaptainTinkham@gmail.com

The changing text of the *Phenomenology of Biocatastrophe* three volume series may be accessed at:

www.biocalert.org

Printed and eBook editions of this three volume series may be purchased from:

amazon.com

Printed editions may also be purchased from:

www.davistownmuseum.org/publications.html

I. Phenomenology of Biocatastrophe

Biohistory

Biocatastrophe is an historical event that occurs over a period of time – not days or years but centuries – as a result of the unique imposition of industrial civilization within a biosphere with a limited capacity to support its ever-growing human communities. Unfortunately, the evolution of biocatastrophe as a natural response of living ecosystems to the impact of human activity is nearing the final decades of its denouement. This climatic moment, ironically, one of disequilibrium rather than equilibrium, is not the apocalyptic final end of the world as described in the last book of the New Testament. Human civilization will continue to exist, but a significant decline in the population of human communities will occur unless radical changes evolve in the interaction between human ecosystems and natural ecosystems. The fundamental question implicit in the biohistory of man's relationship with his environment is how quickly, and utilizing what strategies, will innovating humans adapt to a rapidly deteriorating biosphere with a vulnerable atmospheric water cycle and finite natural resources? As illustrated by the Chernobyl accident and many other anthropogenic cataclysms, nature has an amazing resilience. The ideology of human ecology, as well as the core values of all world religions, offers the hope, or should we say the prayer, that humanity can achieve a sustainable equilibrium with the natural biogeochemical cycles of our Round-World Commons. As with the evolution of the phenomenon of biocatastrophe, this post-apocalypse eventuality will occur over a period of time – probably centuries – if human society is to survive its current predicament.

Human History

Specific historical events, important landmarks in the epic of human history, the significance of which are broadly understood, characterize the phenomenology of biocatastrophe: the first use of atomic weapons at Hiroshima and Nagasaki, the assassination of president Kennedy, the reign of the era of Reaganomics and laissez faire free enterprise finance capitalism, the 911 attack, and the global financial crisis of 2008. Natural events, which have enhanced human impact on natural ecosystems due to increased population densities and sophisticated, if vulnerable, information and transportation technologies also characterize the phenomenology of biocatastrophe: hurricane Katrina, the Haiti earthquake, the Nashville flood, to mention a few recent natural disasters. The volcanic ash-jet engine interface is now a component of every European school child's awareness of the vulnerability of complex technological systems to unexpected disruption by natural events. The unfortunate saga of the Gulf oil spill disaster, a post peak-oil event illustrating our pathetic dependence on obsolete and unsustainable fossil fuel technologies, is a specific historical event with yet to be determined impacts and outcomes. The Gulf spill graphically illustrates the synergism of the ecological, economic, political (note the silence of the Tea Party enthusiasts,) and social components of biocatastrophe. The inescapable visibility of the spreading slicks of "Louisiana crude" provide a dramatic contrast to the nearly invisible presence of nitrogen fertilizer-derived Gulf of Mexico dead zones or the growing human body burdens of anthropogenic chemical fallout – topics not so willingly covered by mass media outlets as the more accessible and politically acceptable stories of destroyed fisheries and fishing communities, polluted Gulf estuaries, wetlands, and beaches, and the gradual extinction of regional cultural and economic institutions.

Flat-World – Round-World Interface

Biocatastrophe is an event that occurs in a context of flat-world information age technologies that allow its thorough scientific documentation and the worldwide dissemination of information about its growing impact. Instant electronic pathways simultaneously create awareness, anxiety, entertainment, and ennui. The conscious selectivity of flat-world media technologies result in stories that can be newsworthy forms of highly lucrative media entertainment. Such riveting and often continuous media coverage (at least until interest wanes, as with the Haiti earthquake) helps perpetuate the ritual of aversion that is a characteristic social and political component of the phenomenology of biocatastrophe. In our journey through the biohistory of the phenomenology of biocatastrophe, our awareness of humankind's current predicament is facilitated by information age technologies that allow flat-world access to all corners of the very round-world biosphere of our Hotel California. We can be tuned in, if we care to listen, to a vast multiplicity of dilemmas, messages, scenarios, health physics events, public opinion, and personal experiences of the highly interconnected communities of our round-world biosphere. We begin our journey with essays (*Volume 1*) about our Hotel California ballroom smorgasbords of chemical fallout delights and the many climatological, social, political, economic, and health physic repercussions of the imposition of human ecosystems upon and within natural ecosystems. In this context, the limiting factor of water, as expressed in the health physics impact of petrochemical ecotoxins, wins the Indianapolis 500 race to oblivion of an imploding industrial society in an overpopulated ecosphere. Other obvious limiting factors to the continuing growth of now globalized industrial society are runners-up in our marathon into the limbo of genobiocide: nonrenewable fossil fuels, greenhouse gas inducement of cataclysmic climate change, worldwide soil depletion, oceanic fisheries degradation, etc. The last *Volume* (3) in the *Phenomenology of Biocatastrophe* series explores the chemistry of anthropogenic ecotoxins, which is a key component of biocatastrophe. It presents a selection of the most important health physics issues that confront humanity in a biosphere in crisis, and also contains a sampler of the databases pertaining to the impact of the race to oblivion of industrial society on biotic media, including humans, and the resulting degradation of living ecosystems. If one finds essays on biocatastrophe boring (*Volume 1*) or our epigrammatic journey through the Hotel California's biosphere ballrooms, staterooms, and passageways, a little overwhelming (*Volume 2*), then the reader should start with *Volume 3* and read the last appendices first. If the ecotoxins in maternal cord blood or breast milk don't alert you to the ongoing health physics impact of the phenomenology of biocatastrophe, perhaps you should turn on your flat-world television and take a break with a few hours of FOX News, or try the cable channels highlighting the many ecotoxic offerings of American consumer culture, now often made in China.

Messages in Many Bottles

This *Volume* (*2*) of our exploration of the phenomenology of biocatastrophe begins with definitions pertaining to natural ecology. Absorption and adsorption are natural processes we learned about in our 6th grade science classes; they are now probably subjects too boring for contemporary classrooms, as well as for some readers of this epigrammatic journey through the simultaneous field of events that constitute the phenomenology of biocatastrophe. As we move further into our lexicon of op eds, news bites, epiphanies, and biocatastrophe potpourri, we can elucidate the phenomenology of biocatastrophe as a series of messages, each of which can be considered a slice of one of many biohistorical pies. We have our round-world constitutional right to slice these pies in any sizes that our ready-access to flat-world information resources helps facilitate. Alternatively, we can also tune out, escaping into the vacuum of information age recreational diversions of our flat-world digital landscapes, partaking in the convenient indulgences of the rituals of a consumer society that is now the main player in the phenomenology of biocatastrophe. For our more educated citizenry, there is no escape from the reality of biocatastrophe; at some point in the mid-21st century, western industrial society, now spreading rapidly to the "developing world," will cross a Rubicon and fully enter the realm of non-sustainability. The certainty of this historical event is now widely recognized and acknowledged by many of the so-called "educated elite"; awareness of this possibility is now also dawning on the mass of unemployed or underemployed workers and the rural poor, who are now the source of a seething mass of social discontent in both developed and developing nations. The convergence of informed awareness, courtesy of our flat-world information technologies, with the social phenomenon of billions of world citizens being excluded from the benefits of a global economy in crisis, constitutes the primary threat to the viability of predatory multinational corporations, equity and hedge funds, and shadow bankers who are the prime movers of the rush to biocatastrophe. The fundamental question is: to what extent, if any, can the multiple environmental, social, political, and economic crises now bedeviling global consumer culture be mitigated by the collective human will to survive the tragedy of a very Round-World Commons? Will our innate capacity for innovation and adaptation and our resilience overcome our collective mad rush to oblivion? How can we overcome or modify the impact of human behavior on the natural cycles of round-world biogeochemistry? Genobiocide, the process of humans destroying the viability of living ecosystems, is now mating with biogenocide, the destruction of human communities by microorganisms as a result of the impact of industrial society on most natural ecosystems. The recent evolution of antibiotic resistant bacteria (ABRB) and the worldwide spread of the AIDS virus are specific examples of the ability of biotic media to undermine human ecosystems. The phenomenology of biocatastrophe is a series of real-world historical events that can be understood scientifically, but which also are spiritual events experienced as noumena – the intuitive recognition of the desecration implicit in the destruction of living ecosystems and the species, including humans, who are their occupants.

World Water Crisis

The fundamental limiting factor for the viability of a global consumer society is the most mundane element in everyday life: water. Hundreds of millions of participants enjoy the fleeting pleasures of the non-sustainable lifestyles of a global consumer society whose social values are typified by the world of petropolitics, predatory multinational corporations, and shadow bankers. Their (our) indulgences in the convivial pleasures of an affluent age of information technology contrast with the worldwide economic challenges and stresses experienced by billions of less fortunate world citizens. The accelerating economic difficulties of western market economies mask the irreversible but often invisible world water crisis that is the key constituent of the phenomenology of biocatastrophe. The ongoing tragedy of the Gulf oil spill may serve the unpleasant function of reminding US and world citizens of the vulnerability of our round-world atmospheric water cycle. As this ongoing salt water biocatastrophe impacts fresh water (estuary) productivity, not to mention the beckoning ambiance of Gulf of Mexico beaches, it will create an enhanced awareness of the consequences of our imposition of non-sustainable human ecosystems on a rapidly deteriorating atmospheric water cycle. An unfortunate irony of the Gulf oil disaster, which is conveniently occurring in the vulnerable backyard of American consumer culture, is that it draws media focus away from the rapidly escalating world water crisis. The highly visible plight of billions of world citizens with increasing water stress and decreasing water quality seldom receives much American media attention. The challenges and costs of providing potable water to the world's growing megacities, including those in China, are an important component of this water crisis. Less visible is the increasing contamination of the atmospheric water cycle, including human tissues

(±50% water,) and maternal cord blood and breast milk (90% water,) with thousands of anthropogenic ecotoxins. Most invisible of all are the links between the spread of industrial and consumer product ecotoxins and the profits of multinational corporations, Wall Street traders, and shadow bankers whose predatory financial manipulations are not limited to subprime mortgage tranches, retirement and pension funds, or the currency fluctuations of the Euro.

The Sociology of Biocatastrophe: Anger Becomes Rage

The key to understanding the phenomenology of biocatastrophe is an awareness of the inevitable link between the tragedy of the Round-World Commons and the reality of the non-sustainability of the western model of finance capitalism. Billions of would-be or once-active participants in a flourishing global consumer society, based in part on the marketing of ecotoxic personal care products and electronic equipment, as well as oversized McMansions, are well aware of what happens when, in the ecology of money ecosystem, the money runs out. The fundamental source of the anger of American Tea Party fascists is the frustrations inherent in rapidly declining economic opportunities for all but a small percentage of American and world citizens. Energy and time-saving flat-world technology, which has provided better paying IT-related jobs for hundreds of millions of rural and urban poor in developing nations, has or will also drastically limit employment opportunities for a vast segment of the middle class in western market economies. The anger of downwardly mobile middle class citizens joins with the seething discontent of the billions of world citizens who cannot afford to participate in a global consumer culture in crisis. Most citizens are acutely aware of the degree of control exercised by the "economic elite" and the multinational corporations that employ them, over local and regional economies. Even more distressing are the lucrative stock market, bond, and shadow banking networks that effectively skim the profits and assets of working people, especially those in developed market economies. Another factor contributing to the economic stresses of the phenomenology of biocatastrophe is the dynamics of an aging population where the worker-retiree ratio, formerly 8 to 1 in 1950 in developed nations, will be less than 2 to 1 by 2030 (Erlanger 2010). An accelerating decline in the public resources available to ensure "social security" is an inescapable consequence of this process; it will also be accompanied by the continued growth of income disparities as the world economy continues to be controlled by a small class of the economically privileged elite. The awareness of the economic inequalities of the modern age extends to the rural poor in many nations, whose social and economic insecurity is the wellspring for the demonstrations and violence seen on the nightly news. It is also the source of the broad appeal of Jihad to non-Christians. Only a few million world citizens are the highly paid beneficiaries of flat-world technologies or the lucky winners in shadow banking and stock market casinos. This economic crisis, only slightly and temporarily mitigated by TARP loans and Euro bailouts, is the social counterpart of the ongoing ecological tragedies of the Round-World Commons – e.g. disappearing potable water supplies and proliferating endocrine disrupting and/or obesity enhancing chemicals. The epigrammatic explorations of this volume of our Hotel California exposition constitute an attempt to raise awareness of the phenomenology of biocatastrophe as an ongoing event in our round-world biosphere-in-crisis that impacts all human communities. This Lexicon is simply the sum total of all messages received or commentary recorded as we listen to flat-world news media while journeying through the labyrinths of the round-world ballrooms, staterooms, and passageways of our Hotel California.

II. Lexicon

Our lexicon is a compilation of definitions, op eds, news bites, and other miscellanea intended to provide some sense of the structure, dynamics, and timing of the illusive phenomena of biocatastrophe. What is it, how does it occur, how do we experience it, what is the probable sequence of its historical genesis? Our lexicon entries pertain to a process of change and decay; they may be tendentious, i.e. subject specific. With respect to our metaphor of the Hotel California, our lexicon labyrinths are its menu of desserts. Also brandies, whiskeys, and aperitifs, not to mention coffees, cappuccino, and Capt. T.'s favorite, Ovaltine/espresso with vanilla silk creamer. Just a little bit of cane juice and palm oil won't hurt anyone. Some after dinner (post-apocalypse) acid reducer would be very helpful. Tums anyone?

Table of Contents

Please note: all citations refer to the extensive bibliographies in *Volume 3*.

Part I - Basic Definitions: Natural Ecology and Human Ecology

Natural Ecology

This section of the biocatastrophe lexicon contains a selection of definitions pertaining to natural ecosystems, the biosphere they inhabit, and the processes and organisms impacted by the anthropogenic ecotoxins created by human activity. These messages, posted randomly in the hallways of the Hotel California, have been collected and inserted in the pockets of the now aging and increasingly absentminded members of Engine Company No. 9 as we wander through the labyrinths of our Hotel California. Less tedious memorandums, bulletins, and op eds follow our footnotes on natural and human ecology.

Abiotic media: Non-living components of the biosphere, i.e. air, fresh and salt water, soil, utilized for monitoring ecotoxin contaminant signals.

Absorption: The process by which airborne gaseous and particulate ecotoxins are incorporated within a solid or liquid, including within atmospheric water vapor and airborne particulates, as a component of organic materials being ingested by micro-consumers such as the bacteria and fungi that occupy the bottom of the food chain, or as endocrine disrupting chemicals (EDCs) within animal and human cellular tissue.

Adsorption: The process by which airborne gaseous and particulate ecotoxins accumulate on the surface of a solid or liquid, including as a component of atmospheric water vapor, on airborne particulates, on organic materials being ingested by micro-consumers, such as bacteria and fungi, or as endocrine disrupting chemicals on the surfaces of cellular tissues.

Advection: "Transport of contaminants due only to the flow of water." (Zogorski 2006, 62).

Aerobic biodegradation: "The breakdown of organic contaminants by microorganisms when oxygen is present. Aerobic biodegradation also is known as aerobic respiration." (Zogorski 2006, 62).

Albedo: The reflection of sunlight by snow pack and ice cover, which helps cool the atmosphere, and the reflection of ultraviolet radiation by the ozone layer in the stratosphere, which protects biotic media at the surface of the earth from its harmful impact.

Ampiphilic: "Ampiphilic describes a molecule combining hydrophilic (water loving) and lipophilic (fat loving) properties." (Miller 2008, 50).

Anabolism: The incorporation of nutrient elements and other complex molecules, including ecotoxins, in living organisms, including bacteria, yeast, molds, green plants, and animals.

Anoxic: The most extreme form of hypoxia, i.e. a near total lack of oxygen, as typified by the bottom layers of oceanic dead zones.

Apoptosis: "Apoptosis is the process of programmed cell death (PCD) that may occur in multicellular organisms... Defective apoptotic processes have been implicated in an extensive variety of diseases. Excessive apoptosis causes atrophy, such as in ischemic damage, whereas an insufficient amount results in uncontrolled cell proliferation, such as cancer." (http://en.wikipedia.org/wiki/Apoptosis).

Atmospheric transport: The primary mechanism for local, regional, and hemispheric distribution of gaseous and particulate ecotoxins. Wet deposition (rain and snowfall events) is the most common vector for absorbed or adsorbed ecotoxins subject to atmospheric water cycle transport; dry deposition is a secondary atmospheric transport mechanism for ecotoxin dispersion.

Autoecology: The study of particular organisms, their traits, and their relationship with other organisms and ecosystems.

Autotrophic ecosystem: An ecosystem in which plants and algae, which use sunlight as their major energy source, are the primary producers.

Autotrophic organisms: Algae and plants that are self-generating, mostly through photosynthesis and the uptake of inorganic nutrients. Autotrophic organisms create organic molecules, such as glucose, from inorganic carbon dioxide. "More than 99% of autotrophic production on earth is through photosynthesis by plants, algae, and certain types of bacteria. Collectively these organisms are termed photoautotrophs." (McGraw-Hill 2002, 57).

Bacteria: Microorganisms that inhabit all areas of the biosphere and play an important role in the biochemistry of living organisms. Along with being biologically significant pathogens, bacteria play an important role in human health by helping the body to harness energy and nutrients from food, keeping the immune system healthy and protecting the human body against other disease-causing bacteria. "There are ten times more bacterial cells in your body than human cells." (Wenner 2007).

Bacterial action: The transformation of elemental nutrients into organic molecules such as proteins, amino acids, and carbohydrates by bacteria.

Bacteriophages: Any of the viruses that infect bacteria

Bacteriostatic: Inhibiting the growth or multiplication of bacteria

Biodegradation: The decomposition of organic molecules by microorganisms, including bacteria, fungi, and other detritivores.

Biodiversity: "…the sum total of all the plants, animals, fungi and microorganisms in the world, or in a particular area; all of their individual variation; and all of the interactions between them. It is the set of living organisms that make up the fabric of the planet Earth and allow it to function as it does, by capturing energy from the sun and using it to drive all of life's processes; by forming communities of organisms that have, through several billion years of life's history on Earth, altered the nature of the atmosphere, the soil and the water of our planet; and by making possible the sustainability of our planet through their life activities now." (Earle 2009, 120).

Biofilm: Communities of microorganisms living on the surface of biotic and abiotic materials.

Biofixation: The biotic process wherein an organism, usually a plant, uses CO_2 as its nutritive source of Carbon, releasing O_2 as waste. It has been suggested that this process in microalgae and other bioengineered autotrophs could be used to curb greenhouse gas emissions.

Biogenic: Produced by living organisms

Biogeochemical cycles: Exchange of materials, including nutrients and ecotoxins, between living and non-living components of the biosphere. Biogeochemical cycles include the global atmospheric water cycle, the oxygen, carbon, nitrogen, potassium, and phosphate cycles, and many other nutrient cycles of lesser importance.

Biogeochemistry: The study of the growth and synthesis of living matter, including the transfer and bioconcentration of organic and inorganic materials in the biosphere, as manifested in the circular pulses of chemical elements moving back and forth between organisms and the environment. "Biogeochemistry is concerned with both the biological uptake and release of nutrients and the transformation of the chemical state of these biologically active substances… The two major processes of biogeochemistry are photosynthesis and respiration… Respiration is the reverse of photosynthesis and involves the oxidation and breakdown of organic matter and the return of nitrogen, phosphorus, and other elements as well as carbon dioxide and water to the environment." (McGraw-Hill 2002, 31). During the processes of respiration and decomposition, bioaccumulated anthropogenic ecotoxins are returned to the biosphere for recycling.

Biogeosphere: The totality of all biotic media in the biosphere, including the nutrient elements of which they are composed.

Biological significance: Having the capacity to cause adverse health effects or genetic modifications in biotic media, including humans.

Biological spectrum of ecology: Protoplasm, cells, tissues, organs, organ systems, organisms, populations, biotic communities, ecosystems, biomes, and biospheres.

Biomagnification: The phenomena of increasing concentration ratios of ecotoxins at higher trophic levels.

Biomass: A living mass of a given population of plants or animals at a specific trophic level, including living communities of biotic media such as forests, corn, or algae that might be used as a renewable energy resource.

Biome: A major community of plants and animals having similar life forms or morphological features and existing under similar environmental conditions; multiple ecosystems sharing one dominant life form (e.g. grasses).

Biosphere: The thin film of living organisms occupying the geochemical environment at the surface of the Earth.

Biotic media: Living organisms, including microorganisms (primary producers), autotrophs, detritivores, herbivores, and carnivores.

Carbohydrates: Organic compounds consisting of carbon, hydrogen, and oxygen, which produce energy and oxygen when oxidized, e.g. glucose, starch, etc.

Carbon cycle: The movement of carbon among atmospheric, terrestrial, and oceanic environments and ecosystems.

Carcinogen: A chemical substance or mixture that induces cancer or increases its incidence.

Catabolism: The metabolism of the complex molecules of living organisms creating simpler molecules and accompanied by the release of energy.

Chemosphere: The atomic elemental constituents of the biosphere, such as carbon, nitrogen, oxygen, and their molecular and photosynthesized components, which range from elemental nutrients to anthropogenic ecotoxins, from living bacteria to human tissues, blood, and excrement. The chemosphere includes the hydrosphere, atmosphere, lithosphere, and all living organisms.

Cholinesterase: The enzyme in avian brain and nerve cells that is a bioindicator of exposure to toxic chemicals.

Climate: "a complex interacting set of components including the oceans, atmosphere, cryosphere, and biosphere." (http://www.essc.psu.edu/).

Climate change: "Climate change affects the warming and acidification of the global ocean, it influences the Earth's surface temperature, the amount, timing and intensity of precipitation, including storms and droughts. On land, these changes affect freshwater availability and quality, surface water run-off and groundwater recharge, and the spread of water-borne disease vectors and it is likely to play and increasing role in driving changes in biodiversity and species' distribution and relative abundance." (UNEP 2007).

Chlorinated solvent: "An organochloride, organochlorine, chlorocarbon, chlorinated hydrocarbon, or chlorinated solvent is an organic compound containing at least one covalently bonded chlorine atom. Their wide structural variety and divergent chemical properties lead to a broad range of applications. Many derivatives are controversial because of the effects of these compounds on the environment."

(http://en.wikipedia.org/wiki/Chlorinated_solvent). "Trichloromethane, tetrachloromethane, trichloroethane, dichloroethene, trichloroethene, and others are often associated with PVC (polyvinyl chloride) production as unwanted by-products." (Gopal 2003, 3).

Cluepid: A family of primarily marine plankton-eating forage fish, including herring, shad, hilsa, and menhaden. The viability of oceanic fisheries is dependent on the sustainability of these lower trophic level species.

Copepod: "A group of small crustaceans found in the sea and nearly every freshwater habitat. Many species are planktonic (drifting in sea waters), but more are benthic (living on the ocean floor), and some continental species may live in limno-terrestrial habitats and other wet terrestrial places, such as swamps, under leaf fall in wet forests, bogs, springs, ephemeral ponds and puddles, damp moss, or water-filled recesses (phytotelmata) of plants such as bromeliads and pitcher plants. Many live underground in marine and freshwater caves, sinkholes, or stream beds. Copepods are sometimes used as bioindicators… Planktonic copepods are important to global ecology and the carbon cycle. They are usually the dominant members of the zooplankton, and are major food organisms for small fish, whales, seabirds and other crustaceans such as krill in the ocean and in fresh water. Some scientists say they form the largest animal biomass on earth." (wikipedia.org; April 8, 2010).

Coprophagia: The characteristic process of the ingestion of fecal pellets by detritivores that have been enriched by microbial activity in other organisms' digestive systems, e.g. fungi. One of many transport vectors for ecotoxins in the environment.

Coral polyps: Coral polyps are tiny animals that are the constituents of coral reefs, which supply algae with nutrients and habitat. The algae make the sugars that feed the coral polyps by photosynthesis and also provide the reefs with their coloration. Elevated ocean water temperatures cause the algae metabolism to create toxins, which then kill the coral polyps resulting in coral bleaching, an indicator of global warming and heat stress.

Cryosphere: The frozen component of the hydrosphere, including glaciers, sea and lake ice, permafrost, and snow pack.

Decomposer bacteria: Bacteria that extract nitrogen and ammonia from organic matter, in contrast to nitrogen-fixing bacteria and algae. Decomposer bacteria are a component of the biofixation-assimilation-cooperation cycle, which benefits all living organisms and plays a role in recycling anthropogenic ecotoxins.

Decomposition: "The process of degrading the energy content of dead tissues and simultaneously releasing the chemicals back into the environment. When those chemicals are released in inorganic forms the process is called mineralization."

(McGraw-Hill 2002, 65). It is during the process of decomposition that anthropogenic ecotoxins are recycled from living organisms to the biogeochemical pathways of the biosphere.

Deep aquifer: A critical but vulnerable source of "a large percentage of water for cities, industries and agriculture." (Odum 1975).

Denitrification: The conversion of nitrate into nitrogen gas by denitrifying bacteria; the nitrogen gas is then released back into the atmosphere.

Desorption: The reverse of the absorption or adsorption processes; macro-consumers, heterotrophs, and autotrophs at higher trophic levels have the capacity to desorb chemical ecotoxins, returning them to biogeochemical cycles where they may again become available for uptake by living organisms.

Detrital absorption: The process by which microbial grazers as micro-consumers ingest organic chemicals including nutrient elements and ecotoxins; a second stage in the primary production of living matter. These detritivores, as grazing microorganisms, are then consumed by other detritivores.

Detritivores: Life forms (fungi, bacteria) that colonize all organic matter, including detritus particles, and digest deceased organisms as well as other organic matter for reincorporation by autotrophic organisms (algae and plants). If chemical ecotoxins have been absorbed or ingested by organisms, including microorganisms, as a component of nutrient element cycling, detritivores become a second stage in the transport of contaminant signals from deceased microorganisms to higher trophic levels of the food web.

Detritus food chain: The incorporation of non-living particulate organic materials, including ecotoxins, into living matter by microorganisms such as bacteria and fungi. When the microorganisms that have ingested organic materials die, they are converted to food for detritus-feeding organisms, which are then consumed by predators within the food web. The detritus food chain is the pathway for anthropogenic ecotoxin uptake by micro-consumers such as saprotrophic bacteria, which transfer ecotoxins to higher trophic levels.

Dietary minerals: Essential nutrients consumed by living organisms, including humans, in their food (calcium, chloride, cobalt, copper, iodine, iron, magnesium, manganese, molybdenum, nickel, phosphorus, potassium, selenium, sodium, sulfur, zinc). The consumption of these nutrients provides a pathway for incorporation of anthropogenic ecotoxins in biotic media.

Domoic acid: A naturally occurring marine toxin in phytoplankton, the production of which is exacerbated by increasing ocean temperatures and terrestrial chemical fallout runoff. Domoic acid (DA) is a worldwide oceanic phenomenon; contaminant pulses of domoic acid are well documented in west coast sea mammals and a Prince Edward Island fishing community. "The algae containing DA is eaten by sea life and passed along up through the food chain. Both shellfish and filter feeding fish (clams, mussels, anchovies, sardines, krill, etc.) can accumulate this toxin without apparent ill effects. However, in marine mammals and humans, DA is a tricarboxylic acid that acts as a neurotoxin." (Channel Islands Marine and Wildlife Institute 2009).

Ecocatastrophe: A synonym for biocatastrophe.

Ecohazard: Any substance or activity that poses a threat to living organisms, their habitat or environment.

Ecology: Ecology is the study of the interrelationship between living organisms and their environment, their biology, and their interaction with the biogeochemical cycles of the biosphere, i.e. the dynamic interaction and interrelationships of the constituents of the biotic environment (microorganisms, plants, and animals) and the abiotic environment (atmosphere, soils, geosphere, cryosphere, and aquatic environments).

Ecosystem: An interactive community of plants, animals, and microorganisms existing within the context of biomes. The biosphere is the totality of the Earth's ecosystems and is constituted by the synergistic interaction of terrestrial, marine, lacustrine, estuarine, and riverine ecosystems, all of which are composed by the life cycles of natural organisms. The Holocene is characterized by the progressive imposition on and destruction of natural ecosystems by human ecosystems. "An ecosystem is the organisms and physical factors in a specific location that are interrelated through the flow of energy and chemicals to form a characteristic trophic structure… The trophic structure of an ecosystem characterizes organisms according to their feeding level and how those feeding relationships of species result in specific patterns of energy flow and chemical cycling." (McGraw-Hill 2002, 57).

Endocrine system: The system of organs in an animal body responsible for the production of chemicals that regulate growth, metabolism, reproduction, thought, and other internal processes. This system can be disrupted by chemicals that mimic those it produces, including synthetic growth hormones, pharmaceuticals designed to correct hormonal deficiencies, and a wide variety of synthetic chemicals that act as endocrine disrupting agents (EDCs).

Endogenous: "The word endogenous means 'proceeding from within'… Endogenous substances are those that originate from within an organism, tissue, or cell… Endogenous processes include the self-sustained circadian rhythms of plants and animals." (http://en.wikipedia.org/wiki/Endogenous).

Entropy: As expressed by the Second Law of Thermodynamics, entropy is the tendency of all matter and energy to seek equilibrium in its simplest possible form. The

14

increasing amount of energy expended by humans to pursue global warfare, increase industrial productivity, or expand consumer product production results in the eventual unavailability of that energy and the natural resources from which it was derived, as well as the accumulation of ecotoxins and greenhouse gases produced by these activities, i.e. entropy.

Enzootic: The presence of native, non-human populations.

Epizootiology: The study of disease and its movement and persistence in non-human animals.

Essential nutrients: Nutrients necessary for bodily function that cannot be produced internally and must therefore be consumed, including minerals (calcium, potassium, iodine, magnesium, selenium, iron, etc.), some vitamins, essential fatty acids (linolenic, linoleic), and essential amino acids (histidine, isoleucine, lysine, methionine, phenylalanine, threonine, tryptophan, valine, arginine).

Exogenous: "Exogenous refers to an action or object coming from outside a system… In biology, an exogenous factor is any material that is present and active in an individual organism or living cell but that originated outside of that organism." (http://en.wikipedia.org/wiki/Exogenous).

Food chain: The multiplicity of energy pyramids that constitute the food webs of the biosphere, including the transport vectors by which nutrient elements, as well as contaminant signals, are bioaccumulated. Ecotoxin biomagnification often occurs at the higher trophic (feeding) levels of the food chain.

Food web: The totality of food chains in an ecosystem.

Gaia hypothesis: The theory that the biosphere, as an integrated series of ecosystems, is a unified, self-regulating, cybernetic system; "an ecological hypothesis proposing that the biosphere and the physical components of the Earth (atmosphere, cryosphere, hydrosphere, and lithosphere) are coupled to form a complex interacting system. This coordinated system of living organisms maintains the climatic and biogeochemical conditions on Earth in a preferred homeostasis. Originally proposed by James Lovelock as the earth feedback hypothesis, it was named at the suggestion of his neighbor William Golding to the Gaia Hypothesis after the Greek supreme goddess of Earth. The hypothesis is frequently described as viewing the Earth as a single organism, wherein life itself controls the physical and chemical conditions of the Earth's surface, atmosphere, and oceans. Lovelock and other supporters of the idea now regard it as a scientific theory, not merely a hypothesis, since it has passed predictive tests." (www.wikipedia.org).

Genetic drift: With respect to natural ecosystems not subject to bioengineering, the process through which evolutionary change takes place due to the inclusion or exclusion of genetic traits through random chance from generation to generation, as opposed to natural selection wherein traits are determined by their favorability towards survival (http://evolution.berkeley.edu/evosite/evo101/IIIDGeneticdrift.shtml). With respect to bioengineering, genetic drift is the windborne transfer of genetically modified seeds or pollen, which may contaminate and thus alter long established, indigenous varieties of plants, as in the case of the accidental release of genetically modified corn pollen.

Global transport: Hemispheric transport of gaseous, liquid, and particulate ecotoxins by evaporation, absorption, solubilization, or adsorption in the global atmospheric water cycle by particulates in the troposphere, or by the biogeochemical cycles of terrestrial or aquatic ecosystems. Global atmospheric water cycle transport is the most efficient of all biogeochemical transport mechanisms.

Greenhouse gasses: The major naturally-occurring greenhouse gasses are water vapor, carbon dioxide, methane, and nitrous oxide. As either naturally-occurring or manmade, greenhouse gases absorb and re-emit infrared radiation, therefore heating up the Earth's atmosphere. Other greenhouse gasses that are manmade include sulfur hexafluoride (SF_6), nitrogen trifluoride (NF_3), trichloroethane (TCA), trichloroethylene (TCE), hydrofluorocarbons, perfluorocarbons, and chlorofluorocarbons. Also see the IPCC list of greenhouse gasses in *Appendix S* in *Volume 3* (http://en.wikipedia.org/wiki/IPCC_list_of_greenhouse_gases).

Hadley cells: Hadley cells are enormous atmospheric vortexes that corkscrew out from the equator. They begin warm and wet, lose moisture as they rise, and eventually descend as dry air creating the world's great deserts. (Cribb 2010, 141).

Hemispheric transport mechanism: The means through which gaseous, particulate, or liquid ecotoxins are spread to any ecosystem in the biosphere from a single or multiple source points and nonpoint source (NPS) phenomena. The major transport mechanisms are: a.) atmospheric - wind, water vapor, and airborne dust and other particulates, b.) terrestrial – within biogeochemical nutrient cycles, including food chain biomagnification, and c.) hydrospheric – as contaminants in lacustrine, riverine, and oceanic ecosystems.

Herbivore: Heterotrophic macro-consumers, such as cattle, sheep, goats, and deer, which ingest autotrophic materials (plants and algae). If the plants being consumed, such as grass in the grazing food chain, are contaminated with radioisotopes or chemical fallout, herbivores become a principle pathway for ecotoxin transport to human consumption.

Herbivore food sources: Fruits, grains, leaves, seeds, roots, algae, and nectar – the vulnerable resources of autotrophic ecosystems

Heterotroph: An organism that consumes other organisms; heterotrophs can be micro or macro-consumers, e.g. algae are micro-consumers, herbivores are macro-consumers; both are dependent on the organic matter produced by autotrophs. Heterotrophs are organisms that require organic carbon sources, e.g. animals that are dependent on the organic materials produced by both autotrophs and other heterotrophs.

Heterotrophic micro-consumers: Decomposers, also called saprophytes, such as bacteria and fungi, that break down and digest dead plant and animal materials, making them available for consumption by other organisms (detritivores). If the nutrient elements within the organic matter being subject to uptake are contaminated by absorbed or adsorbed ecotoxins, heterotrophic micro-consumers become vectors for contaminant pulse movement.

Hydrolysis: The creation of amino acids by enzymes digesting protein matter.

Hydrolyze: To be decomposed by a chemical reaction with water.

Hydrosphere: The totality of aquatic environments, including lacustrine, riverine, estuarine, and oceanic ecosystems.

Hypoxia: Low oxygen in both biotic and abiotic material; hypoxia also references a concentration of less than 2-3 milligrams of oxygen per liter of water (mg/l). Also see anoxia and the definition of hypoxia in the *Human Ecology Definition Section*.

Invertivore: A natural pest controller living midway in the food chain.

Key control mechanisms: The microbial subsystems that regulate the storage and release of nutrients, the behavioral mechanics of predator-prey subsystems, and the limiting factors of elemental nutrient availability, including fresh water and food. Key control mechanisms are often influenced by the impact of pathogens, pests, and GMOs on population densities, and the impact of human ecosystems on natural ecosystems.

Ligand: Any molecule that binds to a specific site on a protein or other molecule.

Lignocelluloses: Crop and timber wastes as well as energy crops, such as napier grass and switchgrass, which can be converted into ethanol or biodiesel by thermo-chemical and enzyme-based processes.

Lithosphere: The solid, outermost layer of the geosphere, the crust and top layer of the Earth's mantle.

Macronutrients: Elements and their compounds needed in large quantities for the survival of biotic media, including humans, e.g. nitrogen, phosphorous, and potassium.

Macrophage: "A large immune cell that envelopes invading pathogens and other foreign material." (Miller 2008, 50).

Metabolism: The chemical process, including the interchange of nutrient elements that occurs within living organisms, which results in growth (anabolism) and the production of energy (catabolism).

Metabolite: A chemical involved in the metabolic process, either as a facilitator or byproduct of living organisms. Also, with respect to abiotic ecotoxins, the chemical byproducts of the decomposition of the molecular structure of a toxin, also called cogeners.

Methane: CH_4, an odorless, gaseous hydrocarbon derived from the decay of plant and animal matter, the chief component of natural gas, and a potent greenhouse gas. CH_4 is a common source material for the production of solvents and Freon, including ecotoxins which are transported by the Earth's biogeochemical cycles. Important sources of methane include coal mines, landfills, waste management facilities, and the release of methane from melting permafrost, which contain methane hydrates in the form of ice-like crystals.

Methanogens: "**Methanogens** are microorganisms that produce methane as a metabolic byproduct in anoxic conditions… They are common in wetlands, where they are responsible for marsh gas." (Wikipedia 2012).

Microbial grazers: The bacteria and fungi that convert organic chemicals into living matter and inadvertently incorporate absorbed or adsorbed ecotoxins into the food chain.

Niche: The special location of a particular organism within an ecosystem hierarchy.

Nitrogen cycle: "The cycling of nitrogen in ecosystems involves microorganisms at several critical steps… The only significant pathway introducing nitrogen into trophic webs is the process of nitrogen fixation in which gaseous, molecular nitrogen is converted into organic nitrogenous compounds by nitrogen-fixing bacteria and blue-green algae… The organic molecules are degraded in the detritus food web by ammonifying bacteria that release ammonium ion or ammonia. Nitrite bacteria convert the ammonia into nitrite, and nitrate bacteria oxidize the nitrogen further into nitrate; those conversions are referred to collectively as nitrification." (McGraw-Hill 2002, 65). Anthropogenic ecotoxins can reenter the food web in association with nitrification in association with plant nutrient uptake.

Osmitadphs: Bacteria, fungi, protozoa, and algae that release organic nutrients extracted from decaying plants or other organisms. These organisms "often excrete hormone-like substances that inhibit or stimulate other biotic compounds of the ecosystems. The function of these lower phyla is to perform any biochemical transformation." (Odum 1975). If the organic nutrients being released are contaminated with absorbed or adsorbed ecotoxins, osmitadphs play a role in the biogeochemical recycling of these ecotoxins.

Oxidation: The reaction of oxygen with other molecules, which results in the creation and release of energy as light, heat, or electricity. Oxidation creates oxides, including CO_2 from the combustion of CO (carbon monoxide) in, for example, the smelting process.

Ozone: A naturally occurring, unstable allotrope of oxygen with three oxygen atoms. As a component of the stratospheric ozone layer, ozone filters out harmful ultraviolet radiation. As a greenhouse gas and an anthropogenic pollutant derived from internal combustion engines and other point sources, ozone contamination lowers air quality by its harmful effects on the respiratory systems of animals and humans.

Ozone layer: A component of the upper troposphere between 15 and 40 km with a concentration range from 2 to 8 parts per million. The depletion of the ozone layer, and the consequential increase in ultraviolet radiation, by volatile, anthropogenic chemical fallout, primarily chlorofluorocarbons, is a major potential threat to human health. Recent limitations on the production and release of chlorofluorocarbons and other ozone layer depleting chemicals by the Montreal Protocol (1987) have resulted in the temporary reduction in the extent of ozone layer depletion.

Pandemic: An outbreak of disease, often a virus, which is spread throughout the world's human communities. An epidemic may be widespread within a country or region; a pandemic affects multiple nations and may be hemispheric in its health physics impact.

Pathogen: An agent or producer of disease; most commonly used to refer to infectious organisms: bacteria (e.g. staphylococcus, clostridium, e-coli), viruses (e.g. Marburg, Legionnaires, Ebola, HIV/AIDS), and fungi (e.g. yeast). The term pathogen is a combination of "patho" (disease) and "gen" (a producer). Pathogen may also refer to a noninfectious agent of disease, such as a chemical.

Pathogenic: The propensity of bacteria and fungi to attack living organisms, producing disease.

Photoautotroph: An organism that utilizes energy from sunlight to produce organic molecules, storing energy in the form of organic matter. During photosynthesis, anthropogenic ecotoxins associated with nutrient elements may be incorporated in the biomass that constitutes living matter.

Photolithoautotroph: Photolithoautotrophs are self-nourishing organisms such as algae and flowering plants that have cells containing chlorophyll and are thus capable of fixing light energy (photo) to build complex organic substances from simple inorganic substances (litho).

Photolysis: The splitting of molecules by means of light energy. In the case of photosynthesis, the water undergoes photolysis, where hydrogen binds to an acceptor and oxygen is released (http://www.biology-online.org/dictionary/Photolysis).

Photosynthesis: The uptake, under the influence of sunlight, of CO_2, H_2O, and elemental nutrients by plants, including algae, to form organic matter and oxygen. Photosynthesis converts light energy into chemical energy by the metabolic pathways of plants using carbon dioxide and water to create oxygen and carbohydrates, with sunlight as the energy source. Typical carbohydrates created by the photosynthetic process are sucrose, glucose, and starch.

Photosynthetic bacteria: Bacteria that are able to carry out photosynthesis by absorbing light and using it to synthesize organic compounds.

Prolactin: Prolactin is a pituitary hormone involved in the stimulation of milk production, salt and water regulation, growth, development, and reproduction.

Protists: Unicellular organisms with an internal digestive system. Also called Protista; many Protists fall into the category of protozoa. Protists include bacteria, molds, fungi, algae, diatoms, foraminifers, and radiolarians.

Radiation: "In physics, radiation describes any process in which energy travels through a medium or through space, ultimately to be absorbed by another body. Non-physicists often associate the word with ionizing radiation (e.g., as occurring in nuclear weapons, nuclear reactors, and radioactive substances), but it can also refer to electromagnetic radiation (i.e., radio waves, infrared light, visible light, ultraviolet light, and X-rays) which can also be ionizing radiation, to acoustic radiation, or to other more obscure processes. What makes it radiation is that the energy *radiates* (i.e., it travels outward in straight lines in all directions) from the source." (wikipedia.org).

Red tide: Red tide is caused by a dinoflagelate phytoplankton, alexandrium fundyense, which as a background species is triggered to produce a naturally occurring toxin that is "1000 times more potent than cyanide… at high enough concentrations the natural pigment in the cells can make the water red, hence the name red tide." (Deese 2010). The phenomenon of red tide is often associated with the seasonal elevated ocean water temperatures and inshore wind that will bring the dinoflagelate phytoplankton near shore. Paralytic shellfish poisoning (PSP) in humans and other mammals can be the result of the red tide phenomenon.

Respiration: The reverse of photosynthesis; the oxidation of organic matter and the return of elemental chemicals, plus CO_2 and H_2O, to the environment.

Saprophyte: Any organisms that live on decayed vegetable matter.

Saprotrophs: Micro-consumers, such as bacteria and fungi, which ingest organic materials, including absorbed or adsorbed ecotoxins, thus providing a pathway for their transport to higher trophic levels of the food chain.

Saprozoic: Obtaining nourishment by absorption of dissolved organic and inorganic materials, as in protozoans and some fungi; feeding on dead or decaying matter.

Second law of thermodynamics: Every occurrence of the transformation of energy is accompanied by a loss in the available amount of energy to do work (entropy). The totality of entropy (spent energy) is continually increasing within the finite ecosystems of the Earth's biosphere.

Specific absorption rate (SAR): "A measure of the rate at which energy is absorbed by the body when exposed to a radio frequency (RF) electromagnetic field. It is defined as the power absorbed per mass of tissue and has units of watts per kilogram. SAR is usually averaged either over the whole body, or over a small sample volume (typically 1 g or 10 g of tissue). The value cited is then the maximum level measured in the body part studied over the stated volume or mass." (wikipedia.org).

Steroid hormones: Estrogens, androgens, progestins, and corticoids. The most important estrogen hormone is estradiol, which regulates reproduction in all vertebrates and some invertebrates, such as lobsters, snails, coral, and sea stars.

Terrestrial ecosystem: The totality of the synergistic interactions of climate, soil, vegetation, and saprobes and animal communities, in a specific terrestrial ecosystem.

Toxicogenomics: The science of the study of the toxicology of ecotoxin biomarkers and their human health risk assessments and economic impact.

Trophic levels: The taxonomic classification of the nutrition feeding levels at which organisms function in the food chain, i.e. bald eagles as well as humans feed at the top of the food chain; fungi and other microorganisms are food producers at the bottom of the food chain. Feeding relationships among organisms living in different trophic levels determine patterns of energy flow, chemical cycling, and ecotoxin biomagnification.

Trophic webs: "There are two different types of trophic webs in ecosystems. The grazing food web is based on the consumption of the tissues of living organisms. The detritus food web is based on the consumption of dead organic material, called detritus in aquatic systems and sometimes referred to as litter in terrestrial ecosystems." (McGraw-Hill 2002, 65). Anthropogenic ecotoxins are recycled to the environment in the form of excretory wastes whenever work is performed by living organisms in any trophic web.

Troposphere: The lowest layer of Earth's atmosphere, bordering the stratosphere, wherein the vast majority of water vapor is found and where weather occurs. Prior to the middle of the 20th century, when nuclear weapons testing and use and space exploration began, this was the principal atmospheric environment for anthropogenic chemical fallout transport.

Vitamins: Essential ingredients for human health, such as biotin, choline, folate, niacin, pantothenic acid, riboflavin, thiamine, vitamins A, B6, B12, C, D, E, and K.

World ecosystem: The totality of ecosystems in the environment, i.e. the biosphere.

Zoonoses: Diseases in microorganisms and animals that have the potential to infect humans.

Zooplankton: A large family of marine organisms ranging in size from small protozoans to jellyfish. They are autotrophic and usually transported by ambient water currents, "Zooplankton feed on bacterioplankton, phytoplankton, other zooplankton, detritus (marine snow) and nektonic organisms. Ecologically important protozoan zooplankton groups include the foraminiferans, radiolarians and dinoflagellates. Their feeding methods include filtering, predation and symbiosis with autotrophic phytoplankton." (Thurman 1997). Zooplankton efficiently digest chemical fallout from atmospheric and terrestrial sources deposited in oceanic ecosystems, which is then biomagnified in higher trophic levels of the marine food web.

Human Ecology

Selected definitions pertaining to the imposition of anthropogenic ecosystems and the ecotoxins they produce upon and within natural ecosystems.

Agrigenetics: For-profit genetic manipulation of crops, which decreases the range of genetic variation and therefore increases crop vulnerability through genetic conformity and gene pool erosion.

Amniocentesis: The sampling of amniotic fluid in the uterus using a hollow needle to screen for developmental abnormalities, now an important tool for evaluating the etiology of autism spectrum disorders as epigenetic in origin, i.e. as a result of malfunctioning genetic switches due to exposure to EDCs or as a result of the cross placental transfer of environmental contaminants in amniotic fluids.

Amniotic fluid: That concentration of blood and nutrients transferred from mother to fetus, now increasingly and universally contaminated (since ±1940) with anthropogenic ecotoxins, such as PBDEs, POPs, EDCs, and other neurotoxins such as methylmercury. (See the *Appendices* in *Volume 3*.)

Anthropogenic: Manmade, i.e. created by human activity.

Anthropogenic radioactivity: The creation of ecotoxic ionizing radioactive substances by human activities, such as weapons production, the generation of electricity by nuclear power, and the production and use of food irradiation equipment and smoke detectors. The most biologically significant radioisotopes are iodine 131 (1/2 T = 8 days), strontium 90 (1/2 T = 30 years), cesium 137 (1/2 T = 30 years), and plutonium 239 (1/2 T = 32,000 years), all characterized by the intensity of their electromagnetic radiation.

Area source: "Area Sources are sources of air pollutants that are diffused over a wide geographical area or are estimated in the aggregate. Area sources include emissions from smokestacks, vents or other point sources, that in and of themselves are insignificant, but in aggregate may comprise important emissions. Examples would be emissions from small dry cleaners or home heating boilers or air toxics volatizing from house painting, chainsaws or lawnmowers." (Casco Bay Estuary Partnership 2008, 8).

Atmospheric water vapor: The most important of all transport vectors for anthropogenic ecotoxins in the biosphere; atmospheric water vapor pathways can vary from localized fallout from specific point sources to the hemispheric wide transfer of ecotoxins from local and regional point sources as well as generic non-point source pollution.

Autism: A developmental disorder characterized by impaired verbal and nonverbal communication and/or social skills, a restricted range of interests, lack of eye contact, resistance to change, obsessive body movements, social isolation, and insensitivity to the feelings of others. The etiology of autism and its increasing incidence may be correlated with the increasing diversity and intensity of chemical fallout ecotoxin contaminant signals in abiotic and biotic media.

Bhopal amendment: A 1990 amendment to the US Clean Air Act that was designed to prevent the release of dangerous toxins. Commonly called the "General Duty Clause" (GDC), this section of the Clean Air Act obligates chemical facilities who handle hazardous chemicals to prevent chemical disasters.

Bioactive: To be biologically active; e.g. as are persistent organic pollutants (POPs), which bioaccumulate in food webs and are often biomagnified in successive trophic levels.

Biocatastrophe: An interrelated series of phenomena characterized by the systematic degradation of natural and human ecosystems by the synergistic interaction of multiple environmental crises. These include overpopulation, cataclysmic climate change, the rapid growth of the unsustainable components of a global market economy, the undermining of sustainable local agriculture by industrial agriculture, chemical fallout ecotoxin pulses from industrial, consumer product, and industrial electronic equipment manufacturing processes and their contamination of the atmospheric water cycle (AWC), and the spread of regional warfare and violent crime due to the availability of high powered weapons and explosives. Biocatastrophe is also characterized by the proliferation of antibiotic resistant bacteria (ABRB), newly emerging and mutating viruses, and genetically modified organisms (GMOs). Biocatastrophe is accompanied by the phenomenon of the psychological denial of its inevitability and the evasion of a discussion of its ecological, social, political, and economic components. These include the growing loss of biodiversity and ecosystem productivity; a gradual decrease in the availability of potable water and whole foods not contaminated with ecotoxins; escalating income, health care, and employment opportunity disparities; increasing social unrest and regional sectarian warfare by growing numbers of the disenfranchised and the unemployed; the increasing debts of western market economies; and the collapse of global consumer society viability and the financial institutions upon which it depends.

Biocide: "A biocide is a pesticide used in non-agricultural applications, mainly as an anti-microbial agent." (Miller 2008, 50).

Bioengineer: Manipulators of human, animal, and plant genomes, primarily for the profit of free market, medical, and industrial agricultural corporations. Future techno-elite bioengineers (TEBs) will invent new versions of the world biosphere, i.e. the marriage of natural and human ecosystems in the form of efficient biomass fuel-producing microorganisms and/or bacteria that will consume plastics and other ecotoxins in landfill dump sites. What will the bacteria do with their ingested ecotoxins? Can the techno-elite bioengineers also create efficient CO_2 ingesting organisms? Who will ensure that a TEB suffering from DOD (defiance obedience disorder) won't invent a viral or bacterial pathogen that wipes out much of humanity?

Biogenic: "Biogenic or background sources refers to the concentrations of Air Toxics that are from natural sources and man-made pollutants that are either still in the air from previous years emissions, or have been emitted outside the inventory area and then transported into the region." (Casco Bay Estuary Partnership 2008, 8).

Biogenocide: As a corollary of biocatastrophe, biogenocide is the phenomenon of microorganisms, especially those not controllable by medical technologies, transporting ecotoxins or acting as pathogens and causing die-offs of biotic media, including human populations. Biogenocide may be equated in a metaphorical sense with the annihilation of human ecosystems by bacteria as a result of the carelessness of non-sustainable industrial society. In contrast, genobiocide is the intentional or accidental extinction by human activity of the biodiversity and productivity of the world ecosystem. Also see genobiocide.

Biologically significant chemical fallout (BSCF): Liquid, gaseous, and particulate emissions produced by the pyrotechnical and petrochemical activities of industrial society that have a deleterious impact on living organisms, including humans. The ecotoxins in anthropogenic chemical fallout play a hidden but major role in the ongoing decline of the biodiversity of most ecosystems.

Biomarkers: Indicators signaling events in biological systems or samples, classified as markers of exposure, markers of effect, and markers of susceptibility. A biomarker is often a xenobiotic substance or its metabolite(s) and may be the contaminating substance itself, present in readily obtainable bodily fluids or excretions (NAS/NRC 1989).

Biomass as commodity: The propensity of human society to use natural ecosystem-derived biomass, both living and dead, as commodities to be harvested and sold in local, regional, and global market economies, e.g. fur, fish, forest, coal, gas, and oil. Fossil fuels are a notable example of a non renewable biomass derived resource that, once depleted, cannot be replenished.

Biopharmaceuticals: One of many specialized fields of bioengineering; the evolution of biopharmaceutical miracle technologies began in 1987 with the distribution of Hamelin's insulin for treatment of diabetes. This pioneering use of recombinant DNA technology also soon included treatments for growth hormone deficiencies, hepatitis, anemia, hemophilia, infertility, and other conditions. The nanotechnologies of biopharmaceuticals in particular, and bioengineering in general, will be key players in helping sustainable human communities survive the age of biocatastrophe. Unfortunately, biopharmaceuticals may also be a future potent weapon for bioterrorists.

Biopiracy: "Exploitation of the traditional plant-based medicines of poor countries without paying for them." (Skidelsky 2008, 60).

Biosophics: The biohistorical study of life on Earth, including the interrelated inorganic and biotic processes of the biosphere.

Biosophy: The biological analysis of human ecology including the study of its capacity for culture, its economic systems as social arrangements, and its characteristic imposition of human ecosystems upon natural ecosystems.

Biosphere as bank account: The for-profit exploitation of the earth's collective resources – fossil fuels, forests, soil, water, phosphate fertilizer deposits, oceanic fisheries, riverine fisheries, lacustrine fisheries, and other ecosystem fauna and flora. Implicit in the depletion of the earth's biomass as bank account is the contraction then collapse of the unsustainable components of growth-based, consumer product-centered, industrial society.

Biotechnology: The use of life forms as tools to be manipulated or re-engineered as saleable commodities. The most primitive forms of human biotechnology include agriculture and fermentation, while modern techniques involve directly altering the genetic code of a specimen rather than relying upon gradual manipulation of the gene pool through natural or artificial selection. "The use of biological systems to create medical, agricultural, and industrial products." (Zachariah 2008, 185).

Black silicon: Silicon wafers treated with brief pulses of intense laser energy in the presence of gaseous sulfur and other chemicals, resulting in silicon that is thousands of times more efficient at absorbing photos and releasing electrons than traditional silicon wafers used in solar installations. "The material has found commercial applications in a number of photodetectors for various imaging and night vision applications. It has potential application for high efficiency solar cells. Black silicon is currently being commercialized by SiOnyx, a Massachusetts-based venture-funded startup company which acquired licensing for the process from Harvard in 2006." (wikipedia.org). Black silicon wafers have the potential to revolutionize solar imaging industries.

Body burden: The total amount of toxic chemicals present in a discreet unit (e.g. gram) of biotic media, including humans, at a given time, e.g. in blood serum, tissues, breast milk, urine, fur, eggs, etc.

Body burden assessments: "Simultaneous measurement of the presence and concentration of chemical compounds or their metabolites in human biospecimens such as blood, urine, breast milk, adipose tissue, hair, saliva and meconium." (http://www.commonweal.org/programs/brc/ 2010).

Boom-bust crop cycles: Variations in crop productivity due to natural events, such as nutrient element unavailability or availability, drought, storm intensity, and/or as a result of the impact of human activities, such as the use of genetically modified crop species or dependence on chemical poisons that damage the environment and ultimately reduce crop yields.

***Bt* crop**: "A crop variety, engineered to contain a gene from the soil bacterium *Bacillus thuringiensis* that produces a toxin effective against one of several insect pests, including European corn borer and corn rootworm." (Gurian-Sherman 2009, 41).

Building-integrated photovoltaics (BIPV): The emerging industry utilizing solar polysilicon manufacturing technologies to produce solar installations integrated in and as building materials, typified by thin film solar technologies for roofing, solar shingles, siding, and paneling that produce solar electricity.

Carbapenems: The most powerful of currently available antibiotics now being used as a last line of defense against drug-resistant superbugs such as *Klebsiella pneumoniae*. News reports as recent as November 17, 2011 report the highest rates of multi-drug-resistant infections are now occurring in England, Greece, Italy, Hungary, as well as in Asia. The increasing presence of drug-resistant superbugs closely correlates with the highest use of antibiotics.

Carbon footprint: "Measure of the impact human activities have on the environment in terms of the amount of green house gases produced, measured in units of carbon dioxide." (Wikipedia).

Carbon nanotubes: Flat sheets of carbon atoms in hexagonal formations that are rolled into cylinders (tubes) and whose length, as measured in micrometers, greatly exceeds their width, measured in nanometers. The cylindrical patterns of carbon nanotubes provide them with unique mechanical and electrical properties. Discovered in the early 1990s, carbon nanotubes are now used to manufacture a wide variety of products, such as electronic components, superplastic machine parts, reinforced concrete, tennis rackets, and other consumer products, and incorporate nanotubes in the form of fibers, films, wires, and circuits. Carbon nanotubes manufacturing processes produce a wide variety of ecotoxins such as polycyclic aromatic hydrocarbons (PAHs) and highly toxic metallic powders. The inadvertent production of carbon nanotubes by ancient ironmongers appears to be a key component of the effectiveness and durability of Damascus and samurai swords.

CdTe quantum dots: Cadmium telluride photovoltaic materials produced as a thin film of semiconductor material applied to glass or plastic; an important emerging technology for producing solar energy, which also produces significant ecotoxins in the form of nanotoxins.

Chemical messengers: "Within an organism, unseen electrical and chemical signals dominate. The pulses and molecules facilitate communication inside a cell, between cells, and among organs and body systems. In people and vertebrate animals, the endocrine, nervous, and immune systems use chemical messengers, such as hormones and enzymes, to regulate body function, control behavior, and integrate hormone, brain, immune, and other body systems." (http://e.hormone.tulane.edu/learning/endocrine-disrupting-chemicals.html 2009).

Chimerica: "Chimerica is a term coined by Niall Ferguson and Moritz Schularick describing the symbiotic relationship between China and the United States." (Wikipedia 2009).

Climate forcing: "Changes in the earth's energy balance that tend to alter global temperatures." (Hansen 2005, 31).

Contaminant pulse: The global or regional presence of biologically significant chemical fallout (BSCF), including persistent organic pollutants (POPs), endocrine disrupting chemicals (EDCs), and other ecotoxins in abiotic and biotic media, often measured in parts per billion (ng/g) or parts per trillion (pg/g), characterized by bioaccumulation and increasing biomagnification in the higher trophic levels of ecosystem food webs.

Contaminant signal: Contaminant pulse-derived traces of ecotoxins emitted from local, regional, and global point sources in pathways to human consumption, as measured in specific samples of abiotic and biotic media, including water, air, human tissue (lipids,) blood, urine, and breast milk.

Corpse pathogens: The bacteria, viruses, and fungi that consume heterotrophic organisms, including macro-consumers, such as dead fish, fowl, and cattle. Corpse pathogens as micro-consumers ingest and recycle corpse ecotoxins, which are then ingested by corpse predators, as exemplified by vultures, feral dogs, etc. making ecotoxins available for recycling as a component of elemental nutrient recycling.

Corpse predators: Any heterotrophic carnivore that consumes the body of any other heterotrophic carnivore. Cannibalism is an example of corpse predation at the highest trophic level. If children ingest ecotoxins, whoever eats the children will ingest the ecotoxins, whether microorganisms, such as bacteria, or feral dogs and rats, such as those feeding on Asian vultures contaminated with Diclofenac and now roaming the landscapes of the post-Apocalypse.

Cracking: Cracking is the thermal decomposition of petroleum above 350 °F to produce heating gasses, fuels, and petrochemicals.

Critical mass: Minimum quantity of a fissionable material, such as plutonium 239, needed to initiate the runaway chain reaction of the splitting of atoms, which is the source of the power of atomic weapons and nuclear power plants.

Critical pollutants: The Agency for Toxic Substances and Disease Registry's (ATSDR) definition of critical pollutants is: "Chemicals that persists in the environment, bioaccumulate in fish and wildlife, and are toxic to humans."

Culture: The political, religious, social, industrial, and military human ecosystems imposed on or within natural ecosystems, including the economic subsystems that produce tools, consumer products, weapons, industrial waste, and chemical fallout, typified by human activities such as industrial production, warfare, and global trade. The characteristics of culture, such as religion, law, and the arts, become subsumed by or subservient to the dominant economic goals of market economies where the ecology of money becomes the supreme social prime mover.

Cypermethrin: "Cypermethrin is a synthetic pyrethroid used as an insecticide in large-scale commercial agricultural applications as well as in consumer products for domestic purposes. It behaves as a fast-acting neurotoxin in insects. It is easily degraded on soil and plants but can be effective for weeks when applied to indoor inert surfaces." (http://en.wikipedia.org/wiki/Cypermethrin).

Debromination: A petrochemical processing strategy that changes the molecular structure of brominated hydrocarbons, making them lipid soluble, but not water soluble.

Deforestation: "The fate of large areas of North American forest after the arrival of colliers producing charcoal fuel for bloomsmiths, blast furnaces, foundries, and other ironmongers. Deforestation also occurred in the Mediterranean, Near Eastern, and European areas in the early Iron Age, and is currently occurring in tropical forests in the twilight years of the pyrotechnic society, due to the systematic exploitation of the Earth's natural resources. Deforestation also occurs as a result of urbanization, biofuel production, and the expansion of industrial and/or agricultural activities and results in increasing atmospheric levels of CO_2 due to the decreasing ability of forests to absorb it during photosynthesis." (Brack 2008).

Deltamethrin: "Deltamethrin products are among some of the most popular and widely used insecticides in the world… While mammalian exposure to deltamethrin is classified as safe, this pesticide is highly toxic to aquatic life, particularly fish, and therefore must be used with extreme caution around water." (http://en.wikipedia.org/wiki/Deltamethrin).

Derivative: "A derivative is typically a bet on a future value of a stock or mortgage or some other underlying asset from which its value is 'derived.' Mortgage insurance – betting that a mortgage won't fail (a 'credit default swap') – for example, is a form of a derivative." (Hartmann 2009, 91).

Developing nations: These are rapidly industrializing nations who have been and will continue to be important players in the viability of globalization and the worldwide spread of market economies and their industrial and consumer product manufacturing. China, India, Indonesia, South Korea, Taiwan, Mexico, Brazil, and Nigeria are the most frequently noted developing nations of economic significance.

Development exposure to ecotoxins: "An embryo and fetus is changing quickly, with rapid cycles of cell division and growth, and massive changes in the patterns of gene activation over time. These cycles provide extensive opportunities for mistakes to occur and be incorporated into the organism. Sometimes these mistakes are mutagenic, sometimes they are based on changes outside the genes… The enzymatic mechanisms that work to detoxify contaminants in adults are not fully developed until after birth… Developmental exposures can cause cancers in children and young adults." (Birnbaum 2003).

Diethylstilbestrol (DES): A synthetic nonsteroid female sex hormone given to women in the 1940s and 50s to prevent miscarriage. DES was banned in 1973 after a link was found to the occurrence of vaginal cancer in daughters of those who had taken DES during pregnancy. DES mimics the physiological effects of estrogen; it has since been linked to other reproductive disorders in both male and female children of users. DES was used as a growth accelerator in beef cattle until 1979.

Dioxins and furans: "Dioxins and furans are chemicals created by heating or burning chlorine-containing compounds in the presence of organic (carbon-containing) materials. Dioxins and furans are among the most toxic chemicals currently known to science. They are known human carcinogens and known endocrine disruptors and result in subtle disruption to infant development at ultra-low (parts per trillion) doses. They also accumulate in fat and breast milk. Major sources are incinerators, power plants, pulp and paper mills, diesel engines, refineries, etc." (http://www.nrdc.org/breastmilk/glossary.asp).

Diseases of civilization: Cholera, plague, influenza, scurvy, beriberi, cardiovascular disease, diabetes, lung cancer, typhoid, malaria.

Dispersants: Chemicals used to dissolve oil and other chemicals after accidental spills. Of particular interest after the Deepwater Horizon disaster in the Gulf of Mexico, millions of gallons of dispersants have been sprayed onto or injected into Gulf waters. The Nalco Company is supplying large quantities of COREXIT 9500, "a simple blend of six well-established, ingredients that biodegrade, do not bioaccumulate and are commonly found in household products." (Nalco; http://www.nalco.com/news-and-events/4255.htm). The exact chemical composition of COREXIT 9500, other dispersants, and many other chemicals is proprietary information, making it difficult to evaluate the toxicity of these chemicals. The dispersants being used in the Gulf of Mexico appear to have a significant impact on organisms of the lower trophic levels of the food chain, especially phytoplankton, protozoa, zooplankton, insect larvae, and fish fry thereby posing a threat to the viability of the food chain and the larger fish and predators feeding at higher trophic levels. Two of many chemicals, which are probable constituents of dispersants, are 2-Butoxyethanol 111-76-2 and 2-Butoxyethanol acetate 112-07-2; while known to be non-bioaccumulative, their carcinogenicity has yet to be evaluated despite the widespread exposure of consumers to these chemicals in household cleaners, furniture finishing, spray painting, and other occupations (ATSDR ToxFAQs 2010). The Nalco website (http://www.nalco.com/news-and-events/4255.htm) provides the following information about COREXIT without specifying the chemicals that are in the dispersant.

- One ingredient is used as a wetting agent in dry gelatin, beverage mixtures, and fruit juice drinks.

- A second ingredient is used in a brand-name dry skin cream and also in a body shampoo.

- A third ingredient is found in a popular brand of baby bath liquid.

- A fourth ingredient is found extensively in cosmetics and is also used as a surface-active agent and emulsifier for agents used in food contact.

- A fifth ingredient is used by a major supplier of brand name household cleaning products for "soap scum" removal.

- A sixth ingredient is used in hand creams and lotions, odorless paints and stain blockers.

This synopsis of the "harmless" ingredients of the dispersant being used in the Gulf oil disaster inadvertently provides a compelling commentary on the wide variety of ecotoxins that are incorporated in widely distributed consumer products, and are now the subject of widespread media scrutiny. Some of these "harmless" ingredients include phthalates, bisphenol-A, alkylphenols, and perfluorinated compounds.

Diversity of chemical signatures: A fundamental characteristic of global ecotoxin contaminant signals.

E. *coli* O157:H7: The most virulent pathogen associated with the industrial agricultural production of meat, especially meat produced in CAFOs. E. *coli* O157: H7 emerged in the 1990s as a new form of E. *coli* associated with the contamination of beef with fecal material in conjunction with the crowded and unsanitary conditions in modern feedlots.

EC$_{50}$: The half maximal effective concentration; the concentration that causes a specific effect in 50% of the test organisms.

Ecological subservience: A key concept in the imposition of human ecosystems upon or within natural ecosystems, a fundamental assumption of environmental fascists and free market economy entrepreneurs, and the rationale for the exploitation of the natural resources of the world biosphere for the benefit of a chosen few.

Ecology 101: A frequently untaught class in human ecology, which would theoretically promote the understanding of the fundamental reality that humans live in one integrated biosphere where bacterial and other microbial ingestion of ecotoxins such as POPs, which often mimic essential nutrients, constitutes a transport vector for the repeated cycling of ecotoxins between organisms and their environment, with a tendency of these ecotoxins to be magnified in upper trophic levels. The ultimate lesson of Ecology

101 is the probability of self-inflicted mass genobiocide of much of humanity due to the careless proliferation of anthropogenic radioactive and petrochemical ecotoxins.

Ecosophy: A definition pertaining to the ideology of human ecology. Coined by Arne Naess, the Norwegian ecologist, ecosophy is one of the first attempts to articulate the dualism of human ecology and natural ecology, asserting the fact that human culture lies outside of nature. Naess' emphasis was on the ideal of "deep ecology," stressing the necessity of human activities that are in harmony with the natural cycles of the biosphere (Naess 1989).

Ecosystem services: Pollination, erosion control, water filtration, and fertility (Patel 2010).

Ecotoxicology: The study of the origins, chemical structures, pathways, and health physics impact of anthropogenic ecotoxins.

Ecotoxin: A chemical toxin manufactured and distributed as a result of human (anthropogenic) military, industrial, and agricultural activities that has a deleterious health impact on biotic media, including on those organisms, such as predatory birds and humans, living at the highest trophic levels and subject to the maximum impact of biomagnified ecotoxins. Naturally occurring ecotoxins such as mercury, arsenic, and lead are remobilized by human activities and follow some of the same transport pathways as anthropogenic ecotoxins.

Ecotoxin metabolite: With respect to ecotoxins, rather than living tissues, the chemical processes by which the molecular structure of an ecotoxin is altered, producing cogeners or new forms of that ecotoxin. Ecotoxin metabolites constitute a significant component of global and regional contaminant pulses and may be more toxic (or less so) than their parent chemical.

Endocrine disrupting chemical (EDC): The wide variety of anthropogenic synthetic chemicals that when adsorbed or ingested by humans or other animals disrupt the functioning of the endocrine system by their capacity to alter genetic codes or to mimic and alter hormone functions. In 1996 the EPA adopted this definition, "an exogenous substance that changes endocrine function and causes adverse effects at the level of the organism, its progeny and/or (sub)populations of organisms."

Environmental endocrine hypothesis: "The assertion that chemicals called 'endocrine disruptors' are interfering with the normal functioning of hormones in animals and humans. The theory is both attractive and troubling – attractive because it offers a unified explanation for a wide variety of ills affecting modern societies; troubling because of its staggering implications for the effects of modern industrial practices." (Krimsky 2000).

Epigenetic modifications: Methylation and histone modifications are "the two main classes of epigenetic modifications – in a diverse set of normal tissues." (Weinhold 2006).

Epigenetics: "The study of the relationship between environmental ecotoxins and genetics." (McKibben 2010), i.e. the science of the study of gene changes and alterations by anthropogenic ecotoxins, including persistent organic pollutants, endocrine disrupting chemicals, heavy metals, natural and anthropogenic radioactivity, and common everyday phenomena such as diesel exhaust, tobacco smoke, and exposure to viruses and bacteria. "Any process that alters gene activity without changing the DNA sequence, and leads to modifications that can be transmitted to daughter cells." (Weinhold 2006).

Epigenomics: The study of the genome wide distribution of epigenetics.

Erythropoietin (EPO): EPO is a naturally occurring hormone that is now genetically engineered and sold by Amgen and Johnson & Johnson (Epogen and Procrit). It stimulates the production of red blood cells and has been used by athletes, such as Lance Armstrong, to improve their performance (Sharp 2012).

Estrogen: A form of steroid hormone commonly used in controlling reproductive activity in animals, including female humans, estrogens move easily across cell membranes and can alter the expression of many genes. Xenoestrogens are industrial-made chemicals that mimic estrogen; as anthropogenic accidentally or deliberately produced ecotoxins, they now widely circulate as contaminant pulses in pathways to human consumption.

Eutrophication: A naturally occurring process in which lakes are turned into bogs and then wetlands by the accumulation of organic matter that reduces the supply, as well as the quantity, of oxygen in the water. Human activities, including nitrogen waste runoff from agriculture, and the generic production of organic wastes by human activities cause rapid eutrophication of lacustrine and riverine environments. The introduction of nitrate and phosphate fertilizer runoff is the primary cause of anthropogenic eutrophication of surface waters, as in the widespread eutrophication occurring in northwestern Europe. Eutrophication may also occur in oceanic ecosystems due to nitrate runoff, as typified by the dead zone in the Gulf of Mexico, whose seasonal maximum now exceeds the size of the state of Connecticut.

External costs: Hidden environmental energy and health costs of western market economy productivity not included in market economy pricing indexes, the collection of which are postponed or assumed by other social entities. A classic example of external costs are the hidden ecological and social costs of producing and consuming hamburgers, typified by the external costs of CAFOs, the health physics impact of ecotoxin-laced hamburger ingestion, and other hidden transportation and production costs.

Factory farm: Factory farm is the colloquial term for concentrated animal feeding operations (CAFOs). "Factory farm production is intensifying worldwide, and rates of new infectious diseases are rising. Of particular concern is the rapid rise of antibiotic-resistant microbes, an inevitable consequence of the widespread use of antibiotics as feed additives in industrial livestock operations... The Infectious Diseases Society of America has declared antibiotic-resistant infections an epidemic in the US, and the FAO recently warned that global industrial meat production poses a serious threat to human health (Sayre 2009)." (*Mother Earth News*; www.motherearthnews.com/Natural-Health/Meat-Poultry-Health-Risk.aspx).

Fifteen most frequently detected VOCs: "The 15 compounds with the largest detection frequency in samples from aquifers, domestic wells, or public wells, based on samples from all wells and an assessment level of 0.2 microgram per liter." (Zogorski 2006, 63).

Fish in fish out (FIFO): The ratio of pounds of fish it takes to produce one pound of farmed fish. Salmon farms have a high FIFO ratio, 5 to 1 in contrast to oysters and some fish, such as catfish and tilapia, which can be grown with a high proportion of a vegetarian diet.

Foliar deposition: An important pathway for the transport of chemical and radioactive fallout to human consumption. Autotrophic foliar deposition results in the ingestion of adsorbed or absorbed ecotoxins by herbivores, as typified by the grass-cow-milk pathway for the transport of strontium 90 to human consumption.

Foliar deposition and absorption/adsorption: Foliar deposition is an important pathway to human consumption of chemical and radioactive fallout, which is deposited on plants and absorbed or adsorbed by autotrophs, as typified by the grass-cow-milk pathway for the transport of strontium 90 to human consumption.

Fractal: Fractals are irregular, fine structures, often of microcosmic dimensions, the shape of which cannot be described in traditional geometric language. Fractals are self-similar replications at all levels of magnification and are present in liquid, gaseous, or crystalline forms. Geological formations, metallurgical microstructures, ice, water vapor, and many other natural phenomena are fractal-like in their patterns of formation and dispersion and are characterized by the difficulty of measuring them using Euclidian geometry. Fractal geometry provides the context for the study of the movement of aerosol-bound ecotoxins on particulates or in vapors in the global atmospheric water cycle.

Fugitive air emissions: Air pollutants not caught by capture systems in photovoltaic and electronic equipment production industries, including nanotoxins resulting from CdTe quantum dot solar PV production, and trichloroethane, acetone, ammonia, and greenhouse gasses, such as nitrogen trifluoride, sulfur hexafluoride (SF_6), trichloroethane (TCA), and trichloroethylene (TCE). Numerous other fugitive airborne ecotoxins are derived from other industrial processes.

Fumigant: "A compound or mixture of compounds that produces a gas, vapor, fumes, or smoke intended to destroy, repel, or control organisms such as insects, bacteria, or rodents. Bromomethane is an example of a fumigant used for large-scale strawberry farming." (Zogorski 2006, 63).

Function of government: The philosophy of radical conservatism asserts that the functions of government relate to the maintenance of public safety. In the age of biocatastrophe, the essential activities of state and federal government are: law enforcement, environmental protection, including biomonitoring, health care and social security for all citizens, equal educational opportunities, and the mitigation of the impact of an out of control shadow banking network on the collective assets of society.

Gene expression: "Gene expression is the process by which information from a gene is used in the synthesis of a functional gene product. These products are often proteins, but in non-protein coding genes such as rRNA genes or tRNA genes, the product is a functional RNA. The process of gene expression is used by all known life - eukaryotes (including multicellular organisms), prokaryotes (bacteria and archaea) and viruses - to generate the macromolecular machinery for life." http://en.wikipedia.org/wiki/Gene_expression).

Genetic engineering (GE): "A technology for inserting genes or regulatory sequences from one organism into the genome of another, thereby allowing the acquired gene to be passed to progeny through reproduction." (Gurian-Sherman 2009, 41). The first examples of biochemical engineering were the use of bacteria to produce alcohol and cheese c. 6000 BC. The genetic engineers of the future may create algae that efficiently consume CO_2, produce fossil fuels in large quantities, or efficiently reduce human populations by the accidental or deliberate production of antibiotic resistant bacteria or other pathogens. Genetically engineered crops have achieved, and have the future potential to achieve, significant gains in world food production. As with the subprime mortgage/credit default swap scam, the initial phase of high productivity (high profits) of genetically engineered crops will likely be followed by the collapse of the long term viability of the genetically engineered monocultures of for-profit corporate/industrial agriculture.

Genetic pollution: The cross pollination of crops and other plants with genetically engineered nanoproducts from biotech manufacturing facilities.

Genetic technology: Biotechnology that alters the genetic constituents of biotic media by gene splicing, embryo engineering, controlled crossbreeding, genetic implants, recombinant DNA technologies, or the use of micro-chemicals. While having significant social, economic, and health physics benefits, genetic technology also poses a risk to biological diversity through the production of genetically altered organisms and persistent microbes, which have the potential to colonize and disrupt ecological niches. The greatest threat of genetic technology may be in the proliferation of genetically modified monocultural seed varieties and their controlled distribution by one or more global corporate monopolies. The consequent reduction of seed varieties available to small farmers is a worldwide threat to crop diversity.

Genobiocide: The intentional or accidental extinction by human activity of the biodiversity and productivity of the biosphere as an integrated world ecosystem, based on the ephemeral illusion of an inalienable right to manipulate, exploit, contaminate, or annihilate the abiotic and biotic media of the biosphere. Genobiocide eventually results in large human population die-offs as the efficient functioning of human ecosystems is undermined by the destruction of natural ecosystems. As a threat to the survival of human communities by intentional actions, genobiocide is genocide practiced on a global scale without the targeting of specific races, ethnic groups, or nations. Also see biogenocide.

Genocide: The deliberate extinction of a race, ethnic group, nation, community, or culture.

Genotoxic: "Damaging to DNA and thereby capable of causing mutations or cancer." (American Heritage Medical Dictionary).

Germ cell mutagenicity: The occurrence of increased mutations in populations of cells and/or organisms due to the presence of ecotoxic or epigenetic agents or substances.

Glass Steagall Act: The Glass Steagall Act regulated a separation between commercial banking and investment activities; it was enacted in 1933 as a response to the stock market collapse and Great Depression. The Act was repealed in 1999 because it was believed to hinder the formation of multi-level, conglomerate financial services companies. Its repeal was a key component of the (2008-?) global financial crisis and a major factor in the emergence of the late twentieth century commercial and shadow banking invention of the marketing of leveraged debt as an asset, as in credit default swaps.

Global containment pulse: The worldwide distribution of ecotoxins through hemispheric transport mechanisms by the biogeochemical cycles of the Earth's biosphere. Rain and snowfall events are the primary mechanism for the tropospheric redistribution of water soluble, volatile and/or lipid soluble, persistent organic pollutants (POPs), endocrine disrupting chemicals (EDCs), and other ecotoxins, which were initially deposited in soils and surface waters. Dry deposition is a secondary transport mechanism for global contaminant pulses. Contaminant pulses are incorporated into biotic media and are subject to further transport, bioaccumulation, and biomagnification in ecosystem food webs. Global trade and transportation systems are tertiary mechanisms for the global transport of ecotoxins already incorporated in food and consumer products.

Global contaminant signals: Ecotoxins derived from specific point sources (i.e. power plants, municipal incinerators, industrial plants,) regional point sources (pesticide application, motor vehicle exhaust,) and non source points (local incinerators, demolition, evaporation, particulate shedding,) which then become incorporated in abiotic and biotic media, often as a component of the global atmospheric water cycle. Global transport mechanisms result in global ecotoxin contaminant signals whose concentration ratios are frequently magnified in the higher trophic levels of the ecosystems that they contaminate.

Global equilibrium: The ideal of global equilibrium articulated in the 1972 Club of Rome report: the maintenance of sustainable economic activities and population levels in a biosphere with finite resources (Meadows 1972).

Global warming potential (GWP): A determination of the efficiency of a molecule as a greenhouse gas and its atmospheric lifetime, measured for a specific time period relative to the same mass of CO_2. The net warming effect of greenhouse gasses, also known as climate forcing, is expressed in SI units as Watts per square meter (W/m^2). The net change in effective global forcing between 1880 and 2003 has been estimated as +1.8W/m^2 (Hansen 2005, 1431).

GMO drift: The accidental release and windborne distribution of genetically altered food crop seeds and pollen; a major threat to the genetic integrity of long established varieties of the same species. Genetic drift undermines the universal right of farmers to plant crops that have not been subject to accidental or deliberate genetic modification and could permanently alter the quality, characteristics, and biodiversity of long established species.

GMO terrorists: Dissident bioengineers that for political or religious reasons create and disseminate dangerous genetically modified organisms such as new variations of antibiotic resistant bacteria, viruses, or terminator genes that are threats to public health or destroy the productivity of industrial agricultural systems.

Goitrogen: "Goitrogens are substances that suppress the function of the thyroid gland by interfering with iodine uptake, which can, as a result, cause an enlargement of the thyroid, i.e., a goiter." (http://en.wikipedia.org/wiki/Goitrogen).

Gram-negative bacteria: "Gram-negative bacteria are bacteria that do not retain crystal violet dye in the Gram staining protocol. In a Gram stain test, a counterstain is added after the crystal violet, coloring all Gram-negative bacteria with a red or pink color. The test itself is useful in classifying two distinct types of bacteria based on the structural differences of their cell walls." (http://en.wikipedia.org/wiki/Gram-negative_bacteria).

Greenhouse gas: Greenhouse gases are those that cause global warming by the process of the radiative forcing of climate change. Water vapor, CO_2, NO_2, and methane are ubiquitous greenhouse gases that occur naturally; human activities such as the burning of fossil fuels, rice farming, and the production of nitrogen fertilizers release sufficient additional quantities of these gases to cause cataclysmic climate change. Other anthropogenic greenhouse gases include a variety of chlorofluorocarbons, perfluorocarbons, hydrofluorocarbons, tetrafluoromethane, hexafluoroethane, and the notorious chemicals used in the production of semiconductors, trichloroethane (TCA) and trichloroethylene (TCE). Mitigation of global warming will necessitate radical limitations on all anthropogenic greenhouse gasses, not just CO_2.

Greenhouse gas effect: The trapping of the infrared energy that would normally be reradiated from Earth to space (Gore 2007).

Greenwashing: Carbon capture and storage.

Half-life: "The time required for the concentration of a compound in a given environmental medium to be reduced to one-half of its original value by one or more processes, such as degradation or transport into another environmental medium." (Zogorski 2006, 63).

Halogenation: Introduction of chlorine, bromine, fluorine, and other halogens into hydrocarbon molecules for the production of persistent organic pollutants (POPs), such as PCBs, PBBs, and PBDEs. Halogenated hydrocarbons have thousands of useful industrial, agricultural, and military applications. Halogenation is a key strategy for the production of the fire-resistant plastic components of the essential everyday equipment of modern global consumer culture, i.e. computers, TVs, cell phones, I-pods, and other handy electronic gadgets of the age of information technology (IT).

Health physics: Having to do with all the ways you get sick and why.

Heat engine: A machine that utilizes heat energy from any source, including fossil fuels, which is then converted to forceful motion (work,) as, for example, in the form of steam under pressure. The internal combustion engine is a heat engine that uses combustion gases to do work. Nuclear reactors are heat engines utilizing steam-powered turbines to generate electricity.

Heat engine fuel shortage: The looming shortage of renewable and nonrenewable biomass-derived fuels faced by global military/industrial/consumer society (GMICS). The fundamental challenge of modern industrial society is to find alternative, renewable energy sources for its industrial heat engines that do not cause cataclysmic climate change and are not ecotoxin point sources.

High fructose corn syrup (HFCS): A ubiquitous component of modern processed foods, HFCS is, in essence, an anthropogenic ecotoxin made from ingredients such as caustic soda that are frequently contaminated with methylmercury, EDCs, and other environmental contaminants.

Holocene: The Age of Man, the Anthropogene, characterized by the accelerating impact of human activities on natural ecosystems; also the name given to the last 10,000 years of earth's history, since the end of the last major glacial epoch and prior to the beginning of the next ice age.

Hormone-disrupting chemicals: Also called endocrine disrupting chemicals (EDCs), hormone-disrupting chemicals are a ubiquitous component of consumer products, including many plastics; pathways to human consumption include ingestion of leached chemicals via food and water, inhalation of toxic dusts, absorption, and, most importantly, the cross-placental transfer of ecotoxins from mother to child through neonatal cord blood or their transport in breast milk.

Human ecology: 1) Human ecology is the study of the impact of anthropogenic activities on the world's biosphere, including the systematic imposition of human ecosystems onto or within natural ecosystems as exemplified by the development of agriculture, the domestication of animals, the evolution of literate civilizations, and the expansion of toolmaking from instruments of manual operation to machine-made machinery and weapons. The ecological impact of human activities, the subject of human ecology as a science, is characterized by the increasingly intensified expenditure of energy in the non-sustainable consumption of renewable and non-renewable natural resources, concomitant increases in the production of anthropogenic ecotoxins by industrial activity, the advent of global warfare, a vulnerable global network of market economies based on industrial/consumer product production, the construction of huge urban and suburban environments based on the accumulation of debt, and the destruction of the biodiversity and productivity of natural ecosystems. 2) An academic definition is "the study of how the distributions and numbers of humans are determined by interactions with conspecific individuals, with members of other species and with the abiotic environment. Human ecology encompasses both the responses of humans to and the effects of humans on, the environment." (McGraw-Hill 2002, 610).

Human Toxome Project: A project sponsored by the Environmental Working Group (EWG), which is dedicated to mapping the progress of the uptake of toxic industrial chemicals in humans (http://www.bodyburden.org/). It is one of the most important sources of information about the proliferation of anthropogenic ecotoxins.

Hydrofracking (hydraulic fracturing): The injection of water under high pressure into underground shale and other rock deposits to facilitate the release of natural gas. A number of chemicals are added to the water being injected into rock deposits that are often two or three miles under the surface. Contamination of surface water supplies and underground aquifers with the ecotoxic injection water has been an ongoing problem at gas recovery locations, many of which, such as those in New York and Pennsylvania, have only recently been discovered.

Hypospadias: An increasingly common birth defect associated with the active ingredient in insect repellants, N,N-Diethyl-meta-toluamide (DEET), resulting in the misalignment of the opening of the male penis (Centre for Research in Environmental Epidemiology, Barcelona, Spain).

Hypoxia: The phenomenon of low oxygen in biotic or abiotic media, particularly important with respect to the process of eutrophication: "Overenrichment of estuaries and coastal waters with nutrients, especially nitrogen, stimulates outbreaks or 'blooms' of algae that consume vital oxygen from the water when they decompose. The effects of hypoxia include fish kills and shellfish bed losses. These losses can have significant detrimental effects on the ecological and economic health and stability of coastal regions." (http://www.esa.org/education/edupdfs/hypoxia.pdf). Riverine and wastewater runoff from terrestrial point sources entering estuarine and oceanic environments containing any significant quantity of oxygen-depleting nitrate fertilizer also contain thousands of other varieties of chemical fallout, often as microcontaminants below the political or chemical limit of detection (LOD), adding to the environmental impact of hypoxia.

Iatrogenic: Treatment-caused diseases.

Iconic events list: A century of exciting human events, particularly in American and European history: the sinking of the Titanic November 1912, the Battle of the Bulge 1916, the Great Depression 1929, Pearl Harbor 1941, the bombing of Dresden 1945, Hiroshima and Nagasaki 1945, assassination of JFK 1963, assassination of Robert Kennedy and Martin Luther King Jr. 1968, Chernobyl 1986, the destruction of the World Trade Towers 2001, the great tsunami of 2004, Hurricane Katrina 2005, the global financial collapse 2008-?, the flooding of Nashville 2010, and the Gulf oil spill 2010. What momentous event will mark the 100th anniversary of the sinking of the Titanic on April 15, 1912 to help us celebrate the ritual of our ongoing, intentional, self-inflicted genobiocide?

40

IGF-1: A naturally occurring growth hormone; its presence can be multiplied by as much as one order of magnitude by the use of the bovine growth hormone rBGH in milk-producing cattle. Excessive levels of IGF-1 have the potential to cause significant adverse health effects including breast, colon and prostate cancer, and neuro-behavioral, endocrine disrupting, and immune functioning disorders. As a result of worldwide concern about bovine growth hormones and elevated levels of IGF-1 in milk consumed by humans, milk produced in America using rBGH is banned in Europe and other countries. IGF-1 also has a wide variety of adverse health effects on cattle, many of which, including mastitis and udder inflammation, require the use of antibiotics. Infection byproducts include excessive levels of pus in milk, which, along with IGF-1, E. coli bacteria, salmonella, lysteria, and a wide variety of anthropogenic ecotoxins also then become microcontaminants in milk.

In silico: An experiment performed in, and data collected from, a computer.

In vitro: An experiment performed in, and data collected from, a test tube or culture.

In vivo: An experiment performed on, and data collected from, a living organism.

Interbank lending ratio (IBLR): High interbank lending rates characterized the global financial panic of 2008-? and were one source of the sudden unavailability of credit, which in turn was the catalyst for the domino effect collapse of multiple commercial, investment, and shadow banking financial institutions.

Kerf: Waste silicon dust created by the sawing of c-Si wafers created in the production of microelectronics and emerging solar photovoltaic (PV) production technologies.

LC_{50}: Lethal concentration at 50%; the concentration that causes mortality in 50% of test organisms within a specified time period.

Life cycle assessment: The evaluation of the ecological impact of any product at every stage of its existence.

Limit of detection (LOD): A reporting unit, e.g. ng/g (parts per billion), below which a reliable measurement of a given ecotoxin cannot be made.

Limiting factors of human ecology: Clean air, clean water, healthy food, energy supplies, health care services, transportation networks, and educational and employment opportunities are all limiting factors for the growth and health of human ecosystems. Money is the mother of all limiting factors for the most massive of all human ecosystems, global military/industrial/consumer society.

Limits to Growth: An important report produced by the Club of Rome in 1972 that postulates the future limits of the population and the growth of global military/industrial/consumer society. The report proposed that a state of global equilibrium could be designed such that sustainable economic activities and population levels could be maintained. The report abstract stated "Our world model was built specifically to investigate five major trends of global concern – accelerating industrialization, rapid population growth, widespread malnutrition, depletion of nonrenewable resources, and a deteriorating environment." The report concludes "We suspect on the basis of present knowledge of the physical constraints of the planet that the growth phase cannot continue for another one hundred years. Again, because of the delays in the system, if the global society waits until those constraints are unmistakably apparent, it will have waited too long." (Meadows 1972). The global financial crisis that became painfully obvious in September of 2008 is a probable signal of the end of the systemic expansion of the global economy with the possible exception of economic growth in the developing countries of China, India, Indonesia, Brazil, and a few other nations for decade or two.

Lockean Paradigm: The concept that the production of wealth by the manipulation of natural resources is an inalienable right of a theoretical social contract.

Low-external-input (LEI): "A farming method that applies agro-ecological principles of timing, crop rotation, and integrated pest management, among others, to control pests and increase production. Unlike organic farming, however, minimal use of synthetic fertilizers and pesticides is allowed." (Gurian-Sherman 2009a, 42).

Maximum contaminant level (MCL): "A USEPA drinking-water standard that is legally enforceable, and that sets the maximum permissible level of a contaminant in water that is delivered to any user of a public water system." (Zogorski 2006, 63).

Mercury (organic and inorganic): "Mercury in elemental form is a silver-colored metal that exists as a thick liquid at room temperature, familiar to most people as the silver liquid inside mercury thermometers. Mercury can chemically combine with other elements to form organic (carbon-containing) and inorganic (not containing carbon) compounds… Organic mercury, predominantly methylmercury, [is] found in foods such as fish, and ethylmercury [is] found in some vaccine preservatives and some antiseptics… Non-elemental forms of inorganic mercury [are] found primarily in batteries, some disinfectants, and some health remedies and creams." (Environmental Protection Agency 2007).

Mesothelioma: A once rare but now common form of lung cancer caused by exposure to asbestos fibers; the subject of extensive litigation with payment demands from injured workers in the hundreds of millions of dollars.

Methane hydrate: A probable trigger for cataclysmic climate change. The partial release of the ten thousand trillion tons of methane hydrate on the ocean seafloor due to increased seafloor heating from global warming will be the penultimate component of catastrophic cataclysmic climate change.

Micelle: "An aggregate of molecules, where in an aqueous solution the hydrophilic (water loving) head regions form a protective barrier around the oil containing hydrophobic (water hating) tail regions in the micelle centre." (Miller 2008, 50).

Microplastics: Small particles of plastic pollution derived from polyethylene glycol, which now pervade the marine environment and are a threat to many marine species. Microplastics often take the form of nurdles, 27 million tons of which were manufactured in the US for a variety of purposes ranging from industrial uses (abrasives, exfoliates) to cosmetics such as Aveno products, body wash, toothpaste, deodorant, nail polish, and eye shadow. Microplastic pollution is also the result of the breakdown of plastic litter, especially in the marine environment, as well as the shedding of synthetic fibers by the domestic clothes washing process. "Nurdles comprise roughly 98% of the beach debris collected in a 2001 Orange county study." (Moore 2002).

Microscopic robots: Created as an offshoot of the production of artificial bacteria by nanotechnologists, microscopic robots may be entirely mechanical or biological-mechanical hybrids constituted by precisely designed molecules and released into the environment after mass production by microscopic machines (http://www.icta.org/template/index.cfm).

Millennium development goals: As delineated in the UN GEO$_4$ publication, these goals are the central pillars of the ideology of human ecology: eradicate extreme poverty and hunger, achieve universal primary education, promote gender equality and empower women, reduce child mortality, improve maternal health, combat HIV/AIDS, malaria and other diseases, insure environmental sustainability, and develop a global partnership for development (UNEP 2007, 428).

Mixture toxicity: The synergism of multiple ecotoxins in biotic media, and especially in mothers and children, is the subject of growing concern among public health professionals. One example among thousands of mixture toxicity issues is the combined effects of endocrine disrupting chemicals, neurotoxins such as methylmercury, alkylphenols, and persistent organic pollutants such as dioxin – furan cogeners, especially as they are present in breast milk and maternal cord blood.

Mobile source: "Mobile Sources are sources of air pollution from internal combustion engines used to propel cars, trucks, trains, buses, airplanes, ATV's, snowmobiles, boats, etc." (Casco Bay Estuary Partnership 2008, 8).

Molar concentration: "A measure of the concentration of a solute in a solution, or of any chemical species in terms of amount of substance in a given volume. A commonly used unit for molar concentration used in chemistry is mol/L." (http://en.wikipedia.org/wiki/Molar_concentration). An example of the use of this reporting unit is a in a *Hypertension* article (Valera 2009) about mercury in the blood of Inuit adults and its relationship to hypertension. The mean mercury level in the study population was 50.2 nmol/L (50.2 10^{-9} molar per liter) or just above 0.5 parts per million per liter.

Mutagenicity: The alteration of a cell's DNA by chemicals or radiation.

Mycoestrogens: Estrogenic substances derived from fungi, the most common source of which derives from stored grain. Zearalenone is the most significant mycoestrogen. "It may be one dietary factor that can reduce the prevalence of breast cancer." (Wikipedia.org 2010). Evaluation of less benign health physics risks are continuing (Kuiper-Goodman 1990).

Nano-bio-sensor: "Nano-sensor that incorporates a biologically active interface, e.g. DNA, proteins etc." (Miller 2008, 50).

Nano-composite: "Materials that are created by mixing nanomaterial fillers into a base material." (Miller 2008, 50).

Nano-sensor: "Nanoscale chemical, biological or physical sensory points or system used to detect and convey information about a given environment, e.g. temperature, pH, location, or the presence of diseased tissue." (Miller 2008, 50).

Nanomaterial: Engineered material that utilizes the unique properties of nanoparticles such as relatively large surface areas, greater chemical reactivity, enhanced biological activity, and amplified catalytic behavior. Nanoparticles thus have a much higher toxicological risk in comparison to larger particles of the same chemical composition due to their ability to penetrate biological membranes and impact cells, tissues, and organs, i.e. greater bioavailability and enhanced bioactivity. "Nanomaterials which measure less than 70 nm can even be taken up by our cells' nuclei, where they can cause major damage." (Senjen 2009, 4).

Nanoproducts: Highly toxic ecotoxins created as byproducts of nanotechnology. The greatest threat to the viability of complex anthropogenic ecosystems is the "massive intrusion of nanoproducts into the food chain." (Odum 1983). While Odum was referring to the proliferation of ecotoxins such as POPs and EDCs at often invisible levels of contamination, the age of information technology (IT) and its plethora of electronic equipment, bioengineering innovations, and pharmaceutical ecotoxins provides the opportunity for the worldwide spread of an additional repertoire of newly developed nanotoxins (measured in parts per billion), picotoxins (parts per trillion), and femtotoxins (parts per quadrillion as in many forms of chemical fallout in rainwater), many of which have not yet been identified.

Nanoscience: "Nanoscience and nanotechnologies involve studying and working with matter on an ultra-small scale. One nanometer is one-millionth of a millimeter and a single human hair is around 80,000 nanometers in width. The technology stretches across the whole spectrum of science, touching medicine, physics, engineering and chemistry." (www.nanotec.org.uk). "Nanoscience studies how the arrangement of atoms and molecules… affects the properties of materials… nanotechnologies have the potential to bring significant benefits to consumers, society, the environment and the economy through a range of applications. Examples include in health care, electronics and IT, food, environment, sport and clothing." (www.defra.gov.uk/environment/nanotech). Buy now, pay later.

Nanotechnology: The deliberate engineering of materials, structures, and systems at the atomic and molecular level. The science of manipulating matter in size gradations ranging from scales approaching those of individual atoms, as in electronics, to those of 1 to 100 nanometers (1 nm = a billionth of a meter) as in the industrial production of new materials as well as superplastic metals. This technology is often used in the fabrication of materials and robots, many of which have already been introduced into human ecosystems with an as-yet-unknown impact on natural ecosystems. Among the many benefits of nanotechnology are the production of superplasticity in metals, as in ultra safe jet aircraft and nuclear power plant turbines; its principle drawback is its potential for producing a wide array of as yet unidentified ecotoxins.

Nanowire: A nanowire is an extremely thin wire with a diameter of a few nanometers (nm) or less.

Neurotoxin: Anthropogenic and naturally-occurring ecotoxins (e.g. dioxins, methylmercury) that have the potential to disrupt the functions of the central nervous system. Neurotoxins in the biogeochemical cycles of the biosphere are often present in amounts below the limit of detection (LOD). Neurotoxins have the propensity to biomagnify in pathways to human consumption. Fetuses, infants and young children are at the greatest risk of neurological damage from exposure to bioaccumulated neurotoxins.

Nitrogen fertilizers: A primary source of anthropogenic nitrous oxide (N_2O), a greenhouse gas 300 times more potent than carbon dioxide, which contributes up to 12% of anthropogenic global warming emissions (Smith 2007). Nitrogen fertilizer runoff is a prime cause of oceanic algae blooms, whose death results in the depletion of the oxygen content of seawater, causing oceanic dead zones. See hypoxia.

Nonpoint source (NPS) pollution: Pollution originating from sources such as farmland runoff, motor vehicle emissions, and other generic, often mobile, sources of anthropogenic ecotoxins. The EPA notes nonpoint source pollution "comes from many diffuse sources," including rainfall or snow melt runoff, which picks up "human-made pollutants… Atmospheric deposition and hydromodification are also sources of nonpoint source pollution." (EPA 1994). Point source pollution is, in contrast, ecotoxins discharged from specific locations, such as coal-fired power plants.

Numerical control: "Numerical control refers to the automation of machine tools that are operated by abstractly programmed commands encoded on a storage medium, as opposed to controlled manually via handwheels or levers, or mechanically operated via cams alone." (Wikipedia 2012).

Obesogens: Obesogens are synthetic chemicals that alter the body's metabolism.

Ocean acidification: "Ocean acidification is the name given to the ongoing decrease in the pH of the Earth's oceans, caused by their uptake of anthropogenic carbon dioxide from the atmosphere. Between 1751 and 1994, surface ocean pH is estimated to have decreased from approximately 8.179 to 8.104 (a change of -0.075)." (http://en.wikidpedia.org/wiki/Ocean_acidification). Ocean acidification affects coral reef sustainability, sea shell thickness, and the viability of a wide variety of other marine life.

Ogallala aquifer: North America's largest aquifer, ranging through eight states, from South Dakota south to Texas and west to Colorado. The most important limiting factor for the future productivity of much of America's croplands; depletion of this resource since 1940 has already led to significant loss of irrigated crop lands in Oklahoma and Texas. Contamination of this aquifer with petrochemicals and other ecotoxins may exacerbate future losses of this essential natural resource.

Organochlorines: Hydrogenated organic chemicals made with halogenated chlorine, also called chlorinated hydrocarbons. Many POPs fall into this category of ecotoxins.

Organohalogens: Persistent, bioaccumulative, hydrogenated organic chemicals, including some pesticides, industrial chemicals, and synthetic products, which as endocrine disrupting chemicals (EDCs), cause thyroid dysfunction, decreased fertility, and birth defects in humans as well as decreased fertility and hatching in fish.

Organoleptic analysis: The traditional use of sensory testing (taste, sight, smell, touch) in the meat and poultry inspection techniques of the Food and Drug Administration, now considered grossly inadequate for evaluating the safety of foods now containing a growing variety of possible bacterial infections and thousands of environmental chemicals.

Osmotic power: "Mix salt water with fresh water and there you have it: instant carbon-neutral energy. The process is called osmotic power, and a company called Statkraft has just opened the world's first osmotic power plant in Norway." (cleantechnica.com).

Ozone precursors: Nitrogen oxides (NOx), carbon monoxide (CO) and volatile organic compounds (VOCs) are called ozone precursors because the majority of tropospheric ozone formation occurs when they react in the atmosphere in the presence of sunlight (wikipedia.org; 2010).

Palimpsest: With respect to the archaeological record of human behavior, a layer of accidental durable remnants (ADR) deposited in natural ecosystems and now present as an archaeological record of human activities. A palimpsest as a record of human activity could be a layer of pottery, stone and iron tools, architectural or industrial debris, or chemical fallout contaminant signals.

Palisaded elite: That five percent of the participants in a global consumer culture with the highest income levels and, therefore, recipients of well over fifty percent of the world's income. The palisaded elite have access to, and can afford, sophisticated medical technologies, advanced educational opportunities, modern transportation systems, and the digital electronic equipment of the age of information technology. They usually live in highly secure condominiums, city apartments, or well guarded estates, hence the designation "palisaded." As a result of growing income disparities and the economic meltdown of 2008-?, the percentage of the number of palisaded elite as a function of total population is now falling in developed countries, in contrast to developing countries, such as China or India, where it is still rising.

Pathways: The transportation vectors of anthropogenic ecotoxins through the biogeochemical cycles of natural and human ecosystems, which result in their bioaccumulation in abiotic and biotic media. The global atmospheric water cycle is the primary pathway for ecotoxin transport by rainfall events into abiotic media (surface water, soils). Ecotoxins are often first incorporated into the food chain by microorganisms during elemental nutrient uptake, and tend to be biomagnified in the higher trophic levels of living ecosystems. Continuing pathway transfer mechanisms include: surface and atmospheric water vapor diffusion of both water and lipid soluble ecotoxins, the ingestion of ecotoxin-laced phytoplankton or algae by fish, foliar deposited herbivore ingestion of chemical/radioactive fallout, the ingestion of contaminated detritivores and other organisms by higher trophic level heterotrophs and then by carnivores such as osprey, swordfish, eagles, and humans, and the direct ingestion, inhalation, or adsorption as dust of a wide variety of synthetic chemicals by living organisms, including humans. Ecotoxin transport in humans can result from the cross placental transfer to the unborn fetus and via contaminated breast milk to infants. An emerging pathway of importance for synthetic chemicals and nanotoxins is their epigenetic impact on gene expression as endocrine disrupting chemicals (EDCs). The soil vapor extraction pathway is one of the most efficient and effective transport vectors for volatile organic compounds (VOCs) and semi-volatile organic compounds (SVOCs) that have been improperly discarded in landfills and industrial sites.

Peak oil: "The term Peak Oil refers to the maximum rate of the production of oil in any area under consideration, recognizing that it is a finite natural resource, subject to depletion." (Campbell 2009). At some stage in post-peak oil production, the production cost of a gallon of oil will exceed its market value. Opinions vary widely about the date of peak oil production, which range from around the year 2002 to an optimistic estimate of 2020. An extensive discussion of peak oil as well as a description of the M. King Hubbert model of peak oil production can be accessed on Wikipedia. "Peak oil is the point in time when the maximum rate of global petroleum extraction is reached, after which the rate of production enters terminal decline." (http://en.wikipedia.org/wiki/Peak_oil).

Perchlorates: Salts based on perchloric acid ($HClO_4$), commonly used in the manufacture of explosives and fuels, especially for military weapons, aircraft, and rockets. Other than weapons production derived nuclear wastes, perchlorates are among the most environmentally significant residues of global warfare. Widespread environmental contamination has been detected in drinking water; a major spill of perchlorates occurred in Nevada in the 1990s, contaminating the water supply of the San Joaquin Valley and resulting in a significant contaminant pulse of perchlorates in all California produce. The compilation of perchlorate inventories as a component of biological monitoring data is restricted by governmental security concerns; as a government sponsored ecotoxin, its production and distribution has yet to be regulated. "Perchlorate's interference with iodine uptake by the thyroid gland can decrease production of thyroid hormones, which are needed for prenatal and postnatal growth and development, as well as for normal metabolism and mental function in the adult." (http://www.cdph.ca.gov/CERTLIC/DRINKINGWATER/Pages/Perchlorate.aspx).

Pesticide: A pesticide is any substance or mixture of substance intended for preventing, destroying, repelling or mitigating any pest. A pesticide may be a chemical substance, biological agent (such as a virus or bacterium), antimicrobial, disinfectant or device used against any pest." (wikipedia.org). Forms of pesticides include fungicides, herbicides, and insecticides, many of which now move through the food webs of the biosphere as ecotoxin contaminant pulses.

Photolysis: Chemical decomposition resulting from exposure to radiant electromagnetic radiation, especially sunlight. Photolysis is one of several mechanisms that allow ecotoxin transformation into metabolites and cogeners.

Pharma crops: A crop containing a transgene that creates a pharmaceutical product.

Pharming: Pharming "refers to the use of genetic engineering to insert genes that code for useful pharmaceuticals into host animals or plants that would otherwise not express those genes. As a consequence, the host animals or plants then make the pharmaceutical product in large quantity, which can then be purified and used as a drug product. Some drug products and nutrients may be able to be delivered directly by eating the plant or drinking the milk. Such technology has the potential to produce large quantities of cheap vaccines, or other important pharmaceutical products such as insulin." (http://en.wikipedia.org/wiki/Pharming_(genetics)).

Phytoestrogens: "Phytoestrogens are naturally occurring chemical constituents of certain plants that may interact with estrogen receptors to produce estrogenic effects. Two major groups of phytoestrogens found in people's diets are isoflavones and lignans." (Centers for Disease Control and Prevention 2005, 285). "Evidence is accruing that phytoestrogens may have protective action against diverse health disorders such as prostate, breast, bowel, and other cancers, cardiovascular disease, brain function disorders, menopausal symptoms and osteoporosis." (http://en.wikipedia.org/wiki/Menopausal; http://en.wikipedia.org/wiki/Phytoestrogen). Phytoestrogen is found in red clover and other plants and may have the capacity to mitigate the impact of some environmental chemicals.

Pleiotropic: "Refers to the multiple effects of a gene, some of which may have agronomic or safety implications. Pleiotropic effects are common in transgenic crops because of the unpredictable interactions between the transgene or transgenic protein and the crop genome, but they may also occur in conventional crop breeding. (Gurian-Sherman 2009a, 42).

Plutocracy: "Rule by and for the rich." (Garfinkle 2011).

Point source: The specific source of an ecotoxin, which then may be subject to local or regional dispersion, or if volatile or highly water soluble, may be distributed globally by hemispheric transport mechanisms as a component of biogeochemical cycling. Specific petrochemical plants or nuclear weapons tests are examples of point sources of anthropogenic (manmade) ecotoxins. See nonpoint source pollution.

Polycarbonate plastic: A common plastic used in baby and water bottles, food can liners, and clear plastic cutlery, polycarbonate plastic contains bisphenol-A (BPA), a biologically significant ecotoxin that mimics the action of human estrogen. Due to leaching from these and many other consumer products, BPA is now a nearly universal chemical body burden ecotoxin in the American population.

Polyculture: Polyculture is the cultivation of multiple crops in the same space; in a marine environment, raising multiple fish species in aquaculture facilities.

Polystyrene: A plastic used in the production of Styrofoam food containers, disposable containers, and opaque plastic cutlery and a significant source of endocrine disrupting chemicals (EDC) such as BPA.

Polyvinyl chloride (PVC): A ubiquitous toxic form of plastic utilized in building materials including water pipes, plastic bottles of all kinds, cling wrap, and a wide variety of other consumer products. Incineration of PVC produces dioxin; several endocrine disrupting chemicals (EDC) leach from PVC products.

Precautionary principle: "When an activity raises threats of harm to human health or the environment, precautionary measures should be taken even if some cause and effect relationships are not fully established scientifically… There are five primary pieces that comprise a precautionary approach: 1.) Alternatives Assessment. An obligation exists to examine a full range of alternatives and select the alternative with the least potential impact on human health and the ecological systems, including the alternative of doing nothing. 2.) Anticipatory Action. There is a duty to take anticipatory action to prevent harm. Government, business, and community groups as well as the general public, share this responsibility. 3.) Right to Know. The community has a right to know complete and accurate information on potential human health and environmental impacts associated with the selection of products, services, operations or plans. The burden to supply this information lies with the proponent, not with the general public. 4.) Full Cost Accounting. When evaluating potential alternatives, there is a duty to consider all the costs including raw materials, manufacturing, transportation, use, cleanup, eventual disposal, and health costs even if such costs are not reflected in the initial price. Short- and long-term time thresholds should be considered when making decisions. 5.) Thoughtful Decision Process. Decisions applying the precautionary principle must be transparent, participatory, democratic, and informed by the best available independent science." (http://www.commonweal.org/programs/precautionary-principle.html).

Prime mover: The physical and mechanical media used in any human effort to transmit power in human ecosystems, e.g. foot, hand, water, wind, steam engine, internal combustion engine, etc. (Hunter 1979).

Propane: A ubiquitous form of fossil fuel; propane is a three carbon alkane gas derived from natural gas or oil processing, which is then frequently compressed into its liquid form.

Pyrotechnology: The use of fire to make glass, terra cotta (pottery), cement (lime), iron, steel, and other metals; the primordial energy source of proto-industrial and industrial society.

Ractopamine: A drug that is used as a feed additive to promote leanness in animals raised for their meat. It has been banned in the European Union, China, and Russia (Wikipedia 2013).

Radical conservatism: The belief that the individual has the responsibility to take care of his family, community, and environment in the context of the innovative resilience of a free enterprise system based on the democratic values of an informed bipartisan nonsectarian citizenry living in a society with minimal income and health care access disparities and sustainable lifestyles, industries, and ecoagriculture.

Radioisotope thermoelectric generator (RTG): Generators powered by the decay heat of radioisotopes. The decay heat is converted to electricity by thermocouples. Plutonium 238 is often used as a power source in US satellites. Uranium and Polonium 210 are often used in USSR spacecraft. Satellite accidents can disperse these radioisotopes over wide areas and, in several instances, hemispherically, as with a Pu_{238} accident during the Cold War. See RADNET: Nuclear Information on the Internet http://www.davistownmuseum.org/cbm/Rad.html.

Reactive nitrogen: The chemically active form of nitrogen used by most organisms including crops; "excess reactive nitrogen, which is mobile in air and water, can escape from a farm and enter the global nitrogen cycle – a complex web in which nitrogen is exchanged between organisms and the physical environment – becoming one of the world's major sources of water and air pollution." (Gurian-Sherman 2009b, 1). Reactive nitrogen is a major cause of anoxic oceanic dead zones. See nitrogen fertilizers.

Reactive oxygen species (ROS): "Very small molecules which are highly reactive due to the presence of unpaired valence shell electrons, includes oxygen ions, free radicals and peroxides. ROS form as a natural byproduct of the normal metabolism of oxygen and have important roles in cell signaling. However, during times of environmental stress ROS levels can increase dramatically and result in significant damage to cell structures (oxidative stress)." (Miller 2008, 50).

Reformulated Gasoline (RFG) Program: "A program applied to an area established under the Clean Air Act Amendments in which gasoline contained 2 percent oxygen by weight year-round to control levels of tropospheric ozone." (Zogorski 2006, 64).

Regulatory capture: The phenomenon of governmental agencies becoming pawns of the industry it is supposed to regulate.

REPO 105: The book keeping scam invented at Lehman Brothers and utilized by the shadow banking network to disguise debt as financial transactions such as the sale of collateralized debt obligations. A variation of REPO 105 was used to disguise the extensive debts of the Enron Corporation as well as the Greek government.

Resilience: "A system's ability to bounce back to a reference state after a disturbance, and the capacity of the system to maintain certain structures and functions despite disturbance. If the resilience is exceeded, collapse can occur." (UNEP 2007, 304). An apt description of a key characteristic of sustainable free enterprise political systems enduring the stress of biocatastrophe.

Root uptake: A secondary pathway for uptake of anthropogenic ecotoxins by autotrophs such as beets, carrots, onions, potatoes, radishes, and other root crops. In a world of chemical and radioactive fallout, homegrown "organic" root vegetables are among the safest and most reliable food sources for sustainable community food production efforts due to their slow uptake of most rainfall event-derived ecotoxins.

Safe Drinking Water Act: Originally passed in December 1974 and amended several times since, it sets maximum contaminant levels for ecotoxins in drinking water. See *Appendix F* and *Appendix T* in *Volume 3*, which describes the "Halliburton loophole" exempting hydrofracking ecotoxins from the Safe Drinking Water Act biomonitoring regulations.

Shiga: "The E. *coli* bacteria that killed dozens of people in Germany over the past month have a highly unusual combination of two traits… One trait was a toxin, called Shiga, that causes severe illness, including bloody diarrhea and, in some patients, kidney failure. The other is the ability of this strain to gather on the surface of an intestinal wall in a dense pattern that looks like a stack of bricks, possibly enhancing the bacteria's ability to pump the toxin into the body." (Kolata 2011, A5).

Solar CO_2 debt: The amount of CO_2 produced in the manufacture of energy saving photovoltaic solar panels and thin film solar cells as a percentage of the CO_2 production avoided by the use of these green technologies.

Solar insolation: The amount of solar energy received by a given area of the earth's surface at a specific latitude, e.g. in southern Europe solar insolation = 1,700 kWh/m²/yr (De Decker 2008).

Speculative real estate super-bubble: The massive national and worldwide overvaluation of real estate assets, the super-bubble of 2007-8, was the legacy of Reaganomics; its collapse triggered the worldwide financial crisis of 2008-?

Stratospheric fallout: Contamination of the biosphere by ecotoxins dispersed in the upper atmosphere typified by nuclear weapons testing and RTG satellite accident derived radioactive fallout. Radioactive and chemical fallout filters down into the troposphere by dry deposition and is often magnified in its intensity by rainfall events in the lower atmosphere, which facilitate its deposition. Stratospheric fallout can also occur as chemical fallout produced from rocket exhaust (e.g. perchlorates) or supersonic aircraft engine emissions.

Sulfur hexafluoride (SF$_6$): A perfluorinated compound and one of the greenhouse gasses listed by the IPCC (International Panel on Climate Control – see *Appendix S* in *Volume 3*). SF$_6$ is widely utilized to clean the reactors used in silicon chip production. "Of the 8,000 tons of SF$_6$ produced per year, most (6,000 tons) is [also] used as a gaseous dielectric medium in the electrical industry, an inert gas for the casting of magnesium, and as an inert filling for windows... According to the Intergovernmental Panel on Climate Change, SF$_6$ is the most potent greenhouse gas that it has evaluated, with a global warming potential of 22,800 times that of CO$_2$ when compared over a 100 year period. However, due to its high density relative to air, SF$_6$ flows to the bottom of the atmosphere which limits its ability to heat the atmosphere." (wikipedia.org).

Synthetic biology: The use of genetic technology to create entirely new organisms by altering or replacing the genetic components of one organism with those from another organism. Synthetic biology utilizing genetically modified organisms, nanotechnology, and nanobiotechnology is now widely used by industrial, agricultural, and food production systems, including the production of nutraceuticals, vitamins, flavors, and a wide variety of cosmetics and other consumer products.

Synthetic lycopene: "Lycopene is a bright red natural colour and powerful antioxidant found in tomatoes and other red fruit. Synthetic lycopene is derived artificially and is increasingly produced at the nanoscale." (Miller 2008, 50).

Synthetic organic biocides: The onset of World War II and the production of DDT marked the beginning of the mature stage of the age of chemical fallout characterized by the attempt to facilitate chemical control of the environment with the production of ecotoxins, the most important of which were synthetic organic biocides. DDT use initially was a highly successful strategy for the control of lice in troops and dislocated human populations.

Synthetic organic chemicals: "In the United States, production and use of synthetic organic chemicals began in earnest in the 1940s World War II era and climbed steadily throughout the second half of the century. Production topped more than 400 billion pounds annually in the late 1990s. Today, compounds used as pesticides, drugs, food and product additives, and industrial chemicals (solvents, lubricants) number more than 80,000." (http://e.hormone.tulane.edu/learning/endocrine-disrupting-chemicals.html 2010).

Technometabolism: The multiple uses of carbon by human ecosystems to create energy, smelt and forge iron and steel, crack (decompose) petroleum into toxic petrochemicals, create industrial agricultural ecosystems, and use organic (carbon-containing molecules) materials to manufacture consumer products of every description. Carbon dioxide emissions, organic chemical wastes (POPs), endocrine disrupting chemicals (EDCs), neurotoxins (methylmercury), nanotoxins, and other environmental contaminants are created in proportion to the technometabolism of modern global military/industrial/consumer society.

Teratogen: An agent, including biotoxins and endocrine disrupting chemicals, capable of causing developmental disorders.

Terminator genes: Genetically modified agricultural crops in which the seeds are infertile and will not produce another crop. Produced by industrial agricultural corporations such as Monsanto, terminator genes ensure corporate control of seed distribution while also posing a major threat to the integrity and survival of all food crops if they are accidentally incorporated in other agricultural systems. There is a risk that terminator genes can be adapted by GMO terrorists to undermine the productivity of industrial agricultural systems.

Total trihalomethane concentration: "The sum of all quantified concentrations for bromodichloromethane, bromoform, chloroform, and dibromochloromethane in a water sample." (Zogorski 2006, 65).

Toxic content profile (TCP): The inventory of ecotoxins in a specific sample of abiotic or biotic media. The TCP is the indicator of the ecotoxic emissions from industrial, petrochemical, or consumer product point sources.

Toxic emissions profile (TEP): The inventory of ecotoxins emitted from a specific sample of abiotic or biotic media during a measured unit of time. The TEP can be useful indicators of the extent and duration of anthropogenic ecotoxin contamination.

Toxic input profile (TIP): The inventory of ecotoxins ingested, inhaled, or absorbed by any media during discreet periods of time. The TIP of maternal cord blood can be used, for example, to evaluate the cross placental transfer of ecotoxins from the mother to the fetus.

Toxic release inventory (TRI) program: An EPA database that contains "information on toxic chemical releases and other waste management activities reported annually by certain covered industry groups as well as federal facilities. This inventory was established under the Emergency Planning and Community Right-to-Know Act of 1986 (EPCRA) and expanded in 1990 by the Pollution Prevention Act (PPA), which required additional data on waste management and source reduction activities be reported under TRI. The TRI-specific sections of these federal laws are section 313 of EPCRA and section 6607 of PPA." (United States Environmental Protection Agency 2006). Only a small proportion of ecotoxins generated by human activity are reported in this inventory, which is dependent on the accuracy of the industrial reports submitted to the EPA. Much larger quantities of ecotoxins are generated by consumer product manufacturing processes, use, and disposal; military activities; overseas industrial production; and accidental release of ecotoxins, and are not a part of this inventory;

nonetheless, it provides an important introduction to the proliferation of anthropogenic ecotoxins by industrial activities. See *Appendix C* in *Volume 3*.

Tranche: The various slices of a collateralized debt obligation consisting of subsidiary investment debts, which have recently included junk bonds, subprime mortgages, and other worthless debt instruments.

Tragedy of the Commons: In its largest context, the propensity of western industrial civilization, now manifested as a rapidly expanding global consumer culture in both developed and developing nations, to use the biosphere as a cesspool for its industrial and chemical wastes. More specifically, a concept penned by Garrett Harden in the journal *Science* in 1968 commenting on the destruction of shared resources by predatory human activity. Harden used the example of cows grazing on communal land. The concept is particularly applicable to the late 20th century spread of ecotoxins, such as PCBs, PBDEs, chlorinated dioxins, and endocrine disrupting chemicals (EDCs) throughout the food webs of the biosphere, culminating in the ever increasing contamination of all biotic media, including amniotic fluid, human breast milk, blood, tissues, and children. The ultimate manifestation of the Tragedy of the Commons is genobiocide.

Transgenic: "Refers to organisms containing genes that have been inserted into their genetic code, usually from other organisms (transgenes), using methods that isolate the transgene from other genes of the donor organism in the laboratory." (Gurian-Sherman 2009a, 43).

Trichlorosilane: A chlorine-derived component of silicon production and the source of silicon tetrachloride, one of the most toxic waste products of the photovoltaic and electronics industry. After distillation to remove impurities, trichlorosilane is heated and reduced with oxygen to produce silane (SiH_4) gas, which is then heated again to make pure molten silicon used to grow monocrystalline and multicrystalline silicon crystals (c-Si). The result is the production of microchips, the prime mover of the age of information technology; large amounts of waste, including greenhouse gasses and ecotoxins, are also produced by the extremely energy intensive manufacturing process where 80% of the initial metallurgical grade silicon is lost.

Tube furnace: A reactor or oven used to grow carbon nanotubes, possibly the most important new source of ecotoxins in the 21st century.

Volatile organic compound (VOC): "An organic chemical that has a high vapor pressure relative to its water solubility. VOCs include components of gasoline, fuel oils, and lubricants, as well as organic solvents, fumigants, some inert ingredients in pesticides, refrigerants, some compounds used in organic synthesis, and some by-products of water chlorination." (Zogorski 2006, 65).

Vulnerability: "Vulnerability is an intrinsic feature of people at risk. It is multidimensional, multidisciplinary, multisectoral and dynamic. It is defined here as a function of exposure, sensitivity to impacts and the ability or lack of ability to cope or adapt. The exposure can be to hazards such as drought, conflict or extreme price fluctuations, and also to underlying socio-economic, institutional and environmental conditions. The impacts not only depend on the exposure, but also on the sensitivity of the specific unit exposed (such as a watershed, island, household, village, city or country) and the ability to cope or adapt... While earlier research tended to regard vulnerable people and communities as victims in the face of environmental and socio-economic risks, more recent work increasingly emphasizes the capacities of different affected groups to anticipate and cope with risks, and the capacities of institutions to build resilience and adapt to change." (UNEP 2007, 304).

Xenobiotic substance: A chemical, often an anthropogenic ecotoxin, that is found in an organism that does not normally produce or contain such a chemical; it can also refer to a chemical that is present in a much higher concentration than is usually found in that media. A xenobiotic substance is usually an indicator of environmental stress or pollution.

Xenoestrogen: Estrogenic compounds produced outside of biotic media, including humans; a class of biologically significant anthropogenic endocrine disrupting chemicals (EDC), most of which have been produced after 1940 as a component of modern consumer products. Bisphenol-A (BPA) is an example of a synthetic endocrine disrupting chemical xenoestrogen. "Xenoestrogens are novel, industrially made compounds, that have estrogenic effects and differ chemically from archiestrogens (ancient, naturally occurring) produced by living organisms." (wikipedia.org 2010). Their human health physics significance lies in their ability to promote biological activity by activating genetic processes after entering the nucleus of a cell.

Part II – The Labyrinths of Biocatastrophe: Op Eds, News Bites, etc.

The following selections – Op Eds, News bites, Aphorisms and Epiphanies, Words and Concepts to Think About, and Questions to be Answered – reflect the voices of multiple commentators or are the results of years of peripatetic wandering through the labyrinths of the Hotel California ballroom of smorgasbord delights. Implicit in these labyrinths is the ultimate Tragedy of the Round-World Commons, the ongoing and accelerating human attainment of collective, self-inflicted, world genobiocide.

Op Eds

After America world: "Who wins and who loses in the after America world?...The winner, or at least the beneficiaries, will be the vultures of the world – in the financial area, for example, the short sellers, the hopers for the worst, who profit only when prices plummet. In dark chaos there will be many corpses to pick clean." (Starobin 2009, 32).

Amerikan Tea Party Taliban: The Amerikan Tea Party Taliban (TPT) is a broad-based reactionary response to the election and policies of Barak Obama, and is indicative of the growing tensions, anger, and social unrest, which characterize the deteriorating viability of western market economies as a result of new electronic age technologies and changing industrial industries. The not so subtly veiled racism of the Tea Party Taliban is expressed in the term "Obamacare," a synonym for the N-word. The rise of the TPT and the diversity and enthusiasm of their membership has distinct parallels with the rise of National Socialism and fascism in Germany and Italy in the 1930s, which was also an era of financial stress and economic instability. The "anti-government" and "anti-intellectualism" and lack of rationality of the Tea Party Taliban and their idealization of an absolutely uncontrolled free enterprise system is analogous to the fascist exultation of the power of the state and the superiority of the Aryan race. Control of political opposition to the TPT and the Republican Party that embodies its values is facilitated by electronic media propaganda as demonstrated by the Fox News network and anonymous political advertisements, as opposed to the use of the Gestapo and SS in Nazi Germany. The ideological fanaticism of the Amerikan Tea Party Taliban is the counterpart of the fanaticism of the religious and social values of the Muslim Taliban. (HS October 28, 2010).

Atmospheric water vapor increase: "The large increase in [atmospheric water vapor] W_0 is primarily due to human-caused increases in [greenhouse gases] GHDs ... and not to solar forcing or to the recovery from the Pinatubo eruption... there is an emerging anthropogenic signal in both the moisture content of the Earth's atmosphere and the cycling of moisture between atmosphere, land and ocean." (Santer 2007).

Big Chem, big harm?: "New research is demonstrating that some common chemicals all around us may be even more harmful than previously thought. It seems that they may damage us in ways that are transmitted generation after generation, imperiling not only us but also our descendants. Yet following the script of Big Tobacco a generation ago, Big Chem has, so far, blocked any serious regulation of these endocrine disruptors, so called because they play havoc with hormones in the body's endocrine system." (Kristof 2012, 11).

Biocatastrophe: The disruptions in humankind's relationship with its natural, technical, electronic, social, economic, and political environments. (JW).

Biosphere as world ecosystem: "The Earth functions as a system: atmosphere, land, water, biodiversity, and human society are all linked in a complex web of interactions and feedbacks. Environment and development challenges are interlinked across thematic, institutional, and geographic boundaries through social and environmental policies… Environmental change and developmental challenges are caused by the same sets of drivers. They include population change, economic processes, scientific and technological innovations, distribution patterns, and cultural, social, political, and institutional processes." (UNEP 2005, 362).

Birthright: Despite the temporary fall in the price of oil, an increasingly smaller percent of higher income families and individuals will be able to afford energy and commodities produced from fossil fuels, the availability of which was once assumed to be the birthright of participants in the global market economy (JW).

Breast milk contamination: "Over 350 contaminants have been found in human breast milk. The message is 'don't stop breast feeding, stop the POPs.' Substances that are persistent and able to bioaccumulate must be phased out, irrespective of their known toxicity, because if effects due to these substances do become evident, it will be too late to correct the situation." (www.oztoxics.org/cmwg/body%20burden/load.html).

Climate change: Graeme Pearman states "Most atmospheric scientists now believe that climate change is going a lot faster and will be far more significant than the public is aware of. (Cribb 2010, 146).

Climax communities: The systematic destruction of the climax communities of stable biomes (forest, savannah, jungle, etc.) is the natural organic result of the imposition of technometabolic human ecosystems on natural ecosystems (JW).

Complex issues: "The problem today is that the American political system seems to have lost its ability to create broad coalitions that solve complex issues… Different policies could quickly and relatively easily move the United States onto a far more stable footing." (Zachariah 2008, 211).

Consumption tax: "Adopt a consumption tax to stop our underground economy and get 10 percent of revenue for all slave labor goods sold in America starting Jan. 1, 2011, with 5 percent going to the states" as stated in an advertisement in the October 31st *The New York Times Book Review* for *America is Now a Socialistic Country* placed by John Rigazio who also states "this book is a must read for all Tea Party members."

Contagion: "Contagion is a fact of life in our interconnected global economy and financial markets… Mr. Rosenberg, former chief economist at Merrill Lynch… estimates that fully half of the mortgage-holding population in the country could be under water by 2011… To save the world from economic collapse, we have transferred the liabilities of the private sector to the public… The next recession will happen more quickly than people think." (Morgenson 2010a).

Default fuel: "Because we have failed to develop alternative energy sources, coal has effectively become the default fuel for the twenty first century." (Roberts 2005).

Digital technology energy savings: "The energy savings realised by digital technology will merely absorb its own growing footprint." (De Decker 2009).

Diminishing returns: "Recent history seems to indicate that we have at least reached declining returns for our reliance on fossil fuels, and possibly for some raw materials. A new energy subsidy is necessary if a declining standard of living and a future global collapse are to be averted." (Tainter 1990, 215).

Eco-nits: Eco-nits are the not-so-silent majority of American consumer society enthusiasts whose political beliefs range from the liberal narcissism of Christian Fascist Free Enterprise Entrepreneurs, to the reactionary environmentalism of the anti-Yucca Mountain advocates and their creed of least-safe radioactive waste storage. All eco-nits are leave-it-in-our-back-yard advocates, promoting the inevitable contamination of the Earth's biosphere with anthropogenic radioactivity and chemical fallout. Eco-nits are the advocates of the environmental version of a credit default swap with respect to both radioactive waste disposal and the proliferation of chemical fallout ecotoxins (HS).

Ecocatastrophe: The disclosure of the fate of man in the Holocene is a result of the gradual unfolding of a "conceptual information commons," (Shulman 2002) which documents the hemispheric transport of anthropogenic chemical fallout in the context of the massive loss of biodiversity and ecosystem productivity, the ultimate nightmare of man in the Holocene and his soap opera, the Tragedy of the World Commons (JW).

Ecology 101 reminder: The organisms most vulnerable to the health physics impact of anthropogenic ecotoxins are those feeding at the highest trophic levels of the food web, e.g. eagles, tuna fish, humans. It wasn't the eagles or the tuna fish that produced these ecotoxins. The ultimate lesson of Ecology 101 is the probability of self-inflicted mass genobiocide of much of humanity as a result of the destruction of the natural ecosystems that sustain human communities (JW).

Ecology of Indebtedness: Once human populations exceed the carrying capacity of the biosphere with its multiplicity of limiting factors, economic growth cannot be sustained through the non-sustainable use, contamination, or destruction of natural resources and ecosystems. Rather, continuing unsustainable production and consumption is only made possible by the increase of world indebtedness needed to finance the predatory activities of a blatantly aggressive global for-profit market economy. The Ecology of Indebtedness insures that only human populations engaged in sustainable lifestyles will survive the next several centuries of biocatastrophe; even they may be victims of the predatory environmental fascists and shadow banking financial engineers responsible for some of the most egregious anthropogenic causes of biocatastrophe (Capt. T.).

Economic cycles: The economic cycles of growth, recession, and depression, which characterize western, and now global, market economies, are a prelude to a sustained global financial crisis. This future crisis, larger and more prolonged than the "near economic meltdown" of 2008, will be rooted in the unsustainable growth of global consumer society and the rapid increase in the total debt of the world's complex banking, financial, and governmental infrastructures in proportion to its net assets. Human ecosystem infrastructure collapse – the loss of industrial productivity, employment opportunities, the integrity of public water supplies, the efficiency of transportation and communication systems, the viability of commercial banking systems, health care availability, reliable public safety – is a manifestation of the larger phenomena of biocatastrophe as the natural result of the historical, systematic imposition of vulnerable human ecosystems based upon the ecology of money upon and within even more vulnerable natural ecosystems (CRP).

Ecotoxin contaminant pulses (buy now, pay later): A fundamental component of biocatastrophe is the inventory release delay of hundreds of billions of tons of irretrievable petrochemical and radioactive wastes into the biogeochemical cycles of the biosphere generated by global military/industrial/consumer society since the beginning of the mature stage of the Age of Chemical Fallout (>1945) (CBM).

EDC pathways: "Because they are everywhere, wildlife and humans face a constant barrage of toxic chemicals. Air, water, dust, and animals carry EDCs [endocrine disrupting chemicals] to every continent. Exposure occurs in every stage of life, from development in egg or womb to adulthood, and any number of sources, including food, air, water, soil, household products, breast milk, and the womb... Although EDCs can affect wildlife, the human health risks that may be associated with these mostly low-level yet constant exposures are still largely unknown and highly controversial." (http://e.hormone.tulane.edu/learning/endocrine-disrupting-chemicals.html 2009).

60

Endocrine disrupting chemicals (EDCs): "Our biosphere is saturated with chemicals that cause functional disorders. Humans are now carrying burdens of both industrial and agricultural chemicals at concentrations at which adverse endocrine, immune, and reproductive effects have been reported in affected wildlife and laboratory animals. There is growing evidence that some of these humans have also been affected as a result of their parent's exposure to endocrine disrupting chemicals." From testimony before the US House of Representatives, 1994, by Theo Colborn. (Krimsky 2000, 62).

Entrepreneurship: "We need health care, financial reform and education reform. But we also need to be thinking just as seriously and urgently about what are the ingredients that foster entrepreneurship – how new businesses are catalyzed, inspired and enabled and how we enlist more people to do that." (Friedman 2010, 9).

Environmental fascism: The essence of environmental fascism is the ethos of using a world biosphere as a cesspool for anthropogenic wastes that have been created as a result of military activities or for-profit industrial or consumer product manufacturing. The cult of money replaces ethnic nationalism and totalitarianism as the principle controlling force of profit-driven corporate consumer culture. The enthusiastic participants in the western model of a global market economy/consumer society share one common attribute: our misconception of being above and beyond ecology, i.e. we have forgotten to remember that we are not exempt from the ecological consequences of human activity within a biosphere of finite natural resources. The subgenre of the Christian Environmental Fascist believes in exemption from the limiting factors of natural and human ecology due to the blessing of Christological innocence – redemption, exemption, keep praying, keep spraying; hope hormones, growth hormones, all for me, none for you, survival of the fittest (HS).

Feeding the world: Globalized industrial agriculture now has the capacity to feed the current world population of seven billion people, but cannot execute this obligation because five billion world citizens have insufficient personal income to fund the petropolitical-dependent production and transportation costs of the foods they need to survive. This disparity will only increase as fossil fuel production becomes insufficient to power resource-devouring industrial agriculture. The only viable future alternative to industrial agriculture is the redevelopment of sustainable ecoagriculture in all agrarian communities (JW).

Fuel efficiency: "The cap and trade system would effectively harness market forces, not regulations, to achieve higher fuel efficiency." (Roberts 2005).

Fundamental integrity: Despite the large chorus of angry Tea Party enthusiasts and other antigovernment activity, President Barak Obama remains a popular figure for the majority of Americans due to the fundamental integrity of his and Michelle's efforts at reform, bipartisan compromise, and leadership on the pressing issues of the day. Particularly appealing is Michelle Obama's concern for the health of children with respect to issues such as a healthy diet, home gardens, and the necessity of mitigating the accelerating levels of childhood obesity (JW).

Genetically engineered crops: "Genetic engineering has been promoted as an important means for dramatically improving the yields of staple food crops, but there is little evidence to support such a claim. In *Failure to Yield*, the Union of Concerned Scientists provides the most comprehensive evaluation to date of more than two decades of U.S. genetic engineering research and commercialization aimed at increasing crop yield. Our analysis shows that despite tremendous effort and expense, genetic engineering has only succeeded in measurably increasing the yield of one major food or livestock feed crop—and this contribution has been small compared with other available methods." (Gurian-Sherman 2009b).

Genetically engineered foods: "Drought-tolerant cassava, insect-resistant cowpeas, fungus-resistant bananas, virus-resistant sweet potatoes and high-yielding pearl millet are just a few examples of genetically engineered foods that could improve the lives of the poor around the globe. For example, researchers in the public domain have been working to engineer sorghum crops that are resistant to both drought and an aggressively parasitic African weed, Striga." (Ronald 2010).

Genobiocide vs. biogenocide: The difference between genobiocide and biogenocide is that genobiocide is the destruction of human and natural ecosystems and their communities by human activity; biogenocide is the destruction of human ecosystems and communities by microorganisms such as bacteria, especially in response to non-sustainable human intrusion into and destruction of natural ecosystems (JW).

Global credit markets: As of January 1, 2009, total global debt is at least ten times the amount of total gross annual global economic output. Innovative American financial engineers led world investors, stockholders, and commercial bankers in the suicidal expansion of leveraged borrowing for investment purposes where debt as an asset (credit default swaps, etc.) supported the illusion of the limitless growth of the western market economy (Capt. T.).

Global nanotechnology: "Current predictions estimate the value of the global nanotechnology industry at $1 trillion by 2015. It also presents many challenges … we must identify and manage potential risks before harm is done to human health or the environment." (www.defra.gov.uk).

Global warfare: The majority of the industrial activities of human cultures are devoted to local, regional, or global warfare at the expense of the formation of sustainable, interdependent human ecosystems. The monetary profits derived from preparations for global and regional warfare are one of the principal driving forces of global military/industrial/consumer society (GMICS) (GP).

Green revolutions: The first Green Revolution used industrial agricultural technologies to feed billions of world citizens, doubling the world population in less than 50 years. By 2010, almost on half of the world population has become dependent upon, but cannot afford, the nonrenewable fossil fuel-based petrochemical laced products that have efficiently and effectively undermined sustainable indigenous agricultural communities. The second Green Revolution will be the inevitable rebirth of sustainable local ecoagriculture without the huge external costs of the first Green Revolution. For many communities there is no other alternative to the mass produced junk foods of a globalized market economy (JW).

Ground water contamination: "The greatest danger of potential disaster is ground water contamination, especially in deep aquifers." (Odum 1983).

Historians: "The historian's task is not to disrupt for the sake of it, but it is to tell what is almost always an uncomfortable story and explain why the discomfort is part of the truth we need to live well and live properly... A well-organized society is one in which we know the truth about ourselves collectively, not one in which we tell pleasant lies about ourselves." (Grimes 2010, 18).

Impact mitigation: Human society does not have the economic means to mitigate the social and environmental impact of biocatastrophe in an age of the unregulated growth of global consumer culture and the prospect of accelerating infrastructure collapse (HGB).

Income disparities: If one would take 100 billion dollars in profits in income from America's top 10,000 most successful bankers, investors, hedge fund and stock market speculators, and corporate elite (a most theoretical proposal – and only a small component of total profits of 2009) and redistribute this money to America's thirty million able bodied unemployed or underemployed workers, that $3,330 per person would make a major impact on the financial status of those millions who can no longer afford many of the basic necessities of a sustainable lifestyle (CRP).

Infrastructure collapse: The anomalous, unplanned, unpredictable consequences of the synergistic interaction of natural events (hurricane, fire, earthquake, volcanic eruption, flood, drought, sea level rise, climate change, antibiotic resistant infections) with human activities (warfare, CO_2 emissions, oil spills, hydrofracking, resource depletion, biologically significant chemical fallout, overpopulation, financial market collapse) occurs in the context of the interwoven technological grids of a pyrotechnic society. The worst-case scenario for infrastructure collapse will result from the interaction of multiple local and regional urban biocatastrophes, worldwide financial market crises, and the global collapse of natural ecosystem biodiversity and productivity (Capt T).

Invisibility vs. visibility: Oil slicks, solid wastes, heavy metals, and smelly liquid streams of chemical pollutants are the visible components of human industrial and consumer product ecosystems. Persistent organic pollutants such as PCBs, PBDEs, dioxins, pesticide and herbicide residues, ubiquitous hormone disrupting chemicals, and the broad array of pharmaceutical waste are the invisible components of the Tragedy of the World Commons (GP).

Joe the hunter: When English fur trappers arrived in New England in the early 17th century, they harvested all the breeding females. New England's beaver population was nearly extinct by 1670. It took three centuries to reestablish beaver communities in selected locations. It was also 300 years between the free enterprise achievements of Joe the hunter and the appearance of Joe Six-pack, Joe the plumber, and other enthusiastic enablers of American corporate consumer culture (HS).

Just power: "The derivation of just power from the consent of the governed depends upon the integrity of the reasoning process through which that consent is given. If the reasoning process is corrupted by money and deception then the consent of the governed is based on false premises, and any power thus derived is inherently counterfeit and unjust. If the consent of the governed is extorted through the manipulation of mass fears or embezzled with claims of divine guidance, democracy is impoverished. If the suspension of reason causes a significant portion of the citizenry to lose confidence in the integrity of the process, democracy can be bankrupted." (Gore 2007, 3).

Late stage debilitation: "Historically, top world economic powers have found 'financialization' a sign of late stage debilitation, marked by excessive debt, great disparity between rich and poor, and unfolding economic decline. Ordinary citizens suffer most." (Phillips 2006, 268).

Legacy of 9/11: A rapid expansion of the national, regional, and local components of the security apparatus of America's narcissistic environmental fascist consumer society occurred in the post-9/11 era of the Bush administration. The irony of the social, political, and military response to 9/11 was the affirmation, confirmation, and expansion of the political importance and military accomplishments of a worldwide terrorist network whose growth was, in part, a response to a predatory global consumer culture that enriches the few while indebting the majority of its participants. The billions of world citizens living on the margins of, or excluded from, participating in the convivial brave new "flat-world" (Friedman 2006) of modern information technology are supplemented by the millions of individuals who, due to sectarian religious beliefs, also oppose the growing control of a world economy by small numbers of predatory multi-national corporations. It is this majority of world citizens who, in fact, provide an additional source of future participants in social unrest and terrorist acts directed at an increasingly vulnerable global consumer society and its many national governments (GP).

Legacy of Ayn Rand: With the help of Allen Greenspan, and under the influence of Reaganomics, America's credit market debts are in excess of fifty trillion dollars in 2009, not including federal government debt of approximately fourteen trillion. This is approximately one hundred and fifty thousand dollars in debt for every citizen of the United States, all for the enrichment of the one percent of the American population that owns 50% of the country's assets, i.e. the benefits of a completely unregulated free enterprise system (JW).

Merck alert: "Even the drug companies admit there may be a problem! An official at Merck was quoted as saying, 'There's no doubt about it, pharmaceuticals are being detected in the environment and there is genuine concern that these compounds, in the small concentrations that they're at, could be causing impacts to human health or to aquatic organisms.'" (*NY Daily News* March 10, 2008; www.nydailynews.com/lifestyle/health/2008/03/10/2008-03-10_pharmaceuticals_found_in_drinking_water.html?page=4).

Miracle nanotechnologies: The advent of the era of nanotechnology as a component of global military/industrial/consumer society (GMICS) is based on the development of miracle nanotechnologies, which provide the opportunity to contaminate the world's biosphere with entirely new forms of ecotoxins that have not yet been identified or monitored by the technoelite communities of bioengineers, scientists, or information technology entrepreneurs that inadvertently produced them (HS).

Nanoparticles: "The potential for ingested non-degradable nanoparticles to cause long term pathological effects in addition to short term toxicity is of great concern." (Miller 2008, 22).

Nanotech risks: The use of self-replicating nanorobots as weapons of mass destruction in global warfare has the potential to create weapons with impacts equal to or greater than the tactical use of virus or bacteria. Nanorobots would be completely immune to the effects of antibiotics and other traditional methods for containing and combating biological threats (NA).

New Age of Global Warfare: We are now experiencing a new stage in the era of global warfare characterized by increasing regional, sectarian conflicts between the sponsors of the predatory global economy and the marginalized masses of participants who are not its beneficiaries. The spread of social unrest to all communities now participating in an unsustainable global market economy and facing the growing impact of food and employment insecurity may be the principal threat faced by the palisaded elite and their state security apparatus (JW).

Nitrogen pollution: "A recent scientific assessment of nine global environmental challenges that may make the earth unfavorable for continued human development identified nitrogen pollution as one of only three – along with climate change and loss of biodiversity – that have already crossed a boundary that could result in disastrous consequences if not corrected." (Gurian-Sherman 2009b, 2).

Oceanic dead zones: The rapid growth in oceanic dead zones due to the impact of industrial agriculture, the proliferation of contaminant pulses (methylmercury, POPs, and EDCs) in many species and geochemical cycles, and systematic overfishing, are the principal causes of the decline in the productivity of oceanic fisheries, one of the most important limiting factors of human ecology and a fundamental component of biocatastrophe (GP).

Optimist: "The complexity of our understanding of microbial diseases is already advancing much faster than the complexity of the microbes themselves. Sooner or later the microbes won't be able to compete." (Johnson 2006, 250).

Organic farming: "[A] study of agriculture in the developing world showed that organic methods were two to three times more productive than conventional methods. Organic crop rotations that were well managed resulted in higher yields than green revolution methods that also included crop rotations. The researchers concluded that organic farming can produce enough food to feed the world without increasing the agricultural land base." (Badgley 2007; www.rodaleinstitute.org).

Patent rights: "The U.S. patent system poses a dual threat to society. First, it encourages the development of pernicious technologies and inventions, even ones likely to cause substantial harm to people and the environment. Second, it allows private ownership of basic resources that should belong to everyone. Making matters worse, the United States is pushing other governments to adopt these policies and make them the global norm." (http://www.icta.org/template/index.cfm; 2010).

Pharma crops: "Pharma crops constitute a class of products that presents higher risks to public health, the environment, trade, and food brands than other products. Most pharma crops are engineered to produce compounds, often at high levels, that are not normally found in food, for example, hormones, vaccines, and plastics. Many of the foreign chemicals are drugs designed to be active in humans, some at very low doses. These drugs are intended ultimately to treat only a small subset of consumers for a particular medical condition and could prove dangerous to people not suffering from these conditions if accidentally ingested… Contamination can occur as a result of cross pollination with food versions of the same crops or accidental mixing of pharma crop seeds with food crop seeds at any number of points in the production chain. This contamination, in turn, could threaten public health as well as the economic well being of the industries that make up the food chain… In summary, pharma food crops grown outdoors pose serious threats to the food supply, public health, and food industries. In our view, only a ban can achieve the necessary standard of complete protection of the food supply." (Union of Concerned Scientists 2008, 12-3).

Pharmaceuticals as ecotoxins: The accurate documentation of the adverse health impact on biotic media, especially humans, of pharmaceuticals and growth hormones released as waste products may be constricted by corporate conflicts of interest, political and religious factors, and the future lack of public resources of national and world governments in crisis (JW).

Plutocrats: "Obsessed with vacuous celebrity, Americans make it easier than ever for plutocrats to sail under the radar. Corporate heavyweights and bankers may be suborning Congress and ripping off 'we the people' left and right, but we're too busy dancing with the stars to notice." (Garfinkle 2011, 15).

Polychlorinated biphenyls (PCBs): PCBs will be entering the biosphere from demolition sites, landfills, incineration, and miscellaneous point and nonpoint sources as long as there are obsolete electrical equipment, old wiring, paints and solvents, and thousands of other products awaiting disposal in old cellars, factories, and urban (or any) environments. Mr. Junk-it and his many trucks are taking a lot more than household rubbish to your local landfills and incinerators (Capt. T.).

Post apocalypse: If the post apocalypse is that era of the age of biocatastrophe that follows the collapse of global consumer culture, it may be argued that in many areas of Africa and the Near East as well as in the United States, whole regions have already entered the post apocalypse due to infrastructure collapse. The benefits of flat earth globalization haven't reached these communities yet and it is unlikely that they will in the future (GP).

Privatization of knowledge: Proprietary holders of knowledge assets control the market for and availability of essential biotechnical knowledge, which could otherwise be used to support sustainable agricultural ecosystems. Under the guise of intellectual property rights, unregulated and un-reviewed nano-biotechnologies with the potential for significant adverse impacts on natural and human ecosystems can be marketed as, or have the status of, commodities in a profit driven global consumer society (JW).

Propaganda value of electronic media: A uniquely American coalition of right-wing politicians, evangelicals, and multi-national corporate interests "gains access to the public through a cabal of pundits, commentators, and 'reporters'—call it the Limbaugh-Hannity-Drudge axis. This fifth column in the fourth estate is made up of propagandists pretending to be journalists. Through multiple overlapping outlets covering radio, television, and the Internet, they relentlessly force-feed the American people right-wing talking points and ultra-conservative dogma disguised as news and infotainment—24 hours a day, 7 days a week, 365 days a year. It is quite a spectacle." (Gore 2007, 66).

Reaganomics: 1981 marks the beginning of the era of Reaganomics and the deregulation of the American commercial banking system. This date also marks the zenith of the construction of mammoth wastelands of suburban blight – whole biomes sacrificed for the free enterprise system - the Stupids do real estate. One eventual result of Reaganomics was the subprime mortgage crisis, the catalyst for the disclosure of a component of the American banking system as one large Ponzi scheme based on the endless extension of credit and accumulation of debt. The lesson of Reaganomics is that the efficiency and growth of a global free enterprise system is constrained by limitations of the ecology of money: world debt now exceeds world gross productivity. Future productivity will decline in proportion to the growth in world (and national) debt (Capt. T).

Risk: "In the early eighties, modern finance began to escape reasonable regulation, including the most important regulation of them all, that of the marketplace… the finance industry eventually posed an untenable risk to the economy – a risk that culminated in the trillions of dollars worth of government bailouts and guarantees." (Gelinas 2009).

Roasting chickens: "A debt crisis may be the US vulnerability through which the chickens of bungled oil policy, radical religion-connected delusion in the Middle East, and excessive domestic borrowing come home to roost." (Phillips 2006).

Sewage sludge: "Sewage sludge includes anything that is flushed, poured, or dumped into our nation's wastewater system–a vast, toxic mix of wastes collected from countless sources, from homes to chemical industries to hospitals. The sludge being spread on our crop fields is a dangerous stew of heavy metals, industrial compounds, viruses, bacteria, drug residues, and radioactive material. In fact, hundreds of people have fallen ill after being exposed to sewage sludge fertilizer–suffering such symptoms as respiratory distress, headaches, nausea, rashes, reproductive complications, cysts, and tumors." (http://truefoodnow.org/campaigns/sewage-sludge/).

Shadow bankers as skinheads: At no point did the US Constitution grant the right to American-Christian-Environmental Fascists to pollute the biogeochemical cycles of the biosphere with POPs and EDCs so they could increase the profitability of their stock portfolios. Money, power, greed, and speed are the core of the ideology of worshipers who don't go to church on Sunday, but still believe in the myth of exemption granted to the chosen few. Keep on waving, keep on paving (CRP).

Shadow banking crisis: The recent filing by the SEC against Goldman Sachs is the beginning of the smelting, then melting, of the iron curtains of the shadow banking network and its ability to skim the accumulated wealth of pension funds, stock market portfolios, and investment funds. The predatory activities of Wall Street hedge funds and the trade in collateralized debt obligations (CDOs) are a threat to the viability of a free market economy (TH).

Societal collapse: Traditional causes of societal collapse are warfare, invasions, crop failures, disease, and environmental degradation. Modern technological society, in part based on the productivity of information analysts and producers, in addition to industrial, commercial, and financial activity, faces a new threat in that economic investments in education, technological innovation, energy, and environmental remediation are producing diminishing returns, i.e. "declining marginal returns on added investments in complexity." (Tainter 1990). A fundamental threat to highly technometabolic global consumer society is thus the diminishing returns on investments in education, technology, and environmental remediation in the context of a mythical, ever-growing, global consumer culture (JW).

Solar CO_2 emissions: "It should be realized that solar panels first raise the amount of greenhouse gasses before they help lowering them. If the world would embark on a giant deployment of solar energy, the first result would be massive amounts of extra greenhouse gasses, due to the production of the cells." (De Decker 2008).

Solar PV hazardous materials: "Silicon-based solar PV production involves many of the same materials as the microelectronics industry and therefore presents many of the same hazards. At the same time, emerging thin-film and nanotech-based cells pose unknown health and environmental dangers." (Silicon Valley Toxics Coalition 2009).

Sulfur hexafluoride (SF6): "One ton of sulfur hexafluoride has the greenhouse effect equivalent of 25,000 tons of CO_2. It is imperative that a replacement for sulfur hexafluoride be found, because accidental or fugitive emissions will greatly undermine the greenhouse gas reductions gained by the use of solar power." (Silicon Valley Toxics Coalition 2009).

Survival: There remains the possibility that selected communities and cultures with sustainable economies and ecologies may survive the collapse of global military/industrial/consumer society (JW).

Synthetic halogenated chemicals: "Dioxin alters multiple endocrine systems, and its effects on the developing breast involve delayed proliferation and differentiation of the mammary gland, as well as an elongation of the window of sensitivity to potential carcinogens." (Birnbaum 2003).

Tea Party Taliban: The right to pursue happiness is the fundamental communal commandment of the American Tea Party Taliban (GP).

The ultimate narcissist, excerpts from Beck on common sense: "America is good… I believe in God… My spouse and I are the ultimate authority, not the government… I have a right to life, liberty, and the pursuit of happiness, but there is no guarantee of equal results… I work hard for what I have and I will share it with who I want to. Government cannot force me to be charitable… The government works for me. I do not answer to them; they answer to me." (Beck 2009, 109).

Time is running out: "Living standards are sinking in the United States, and there is no coherent vision on plan for reversing that ominous trend over the long term… the glory days of the go-go American economic empire will fade like the triumph of an aging Hollywood star – steps… need to be taken to remake an economy that has been thrown completely out of whack by frantic, debt-driven consumption, speculation bubbles, exotic financial instruments etc." (Herbert 2010).

Toxic assets: In a world of ongoing global financial collapse, we all become subprime assets. Even worse, in a biosphere of invisible ecotoxin contaminant pulses, adults and children are each other's toxic assets (GP).

Uh-huh: There's nothing wrong with the global consumer society that the free enterprise system won't fix (TPF).

Unanticipated downsides: Flat-world technology now includes the use of growth hormones and genetically modified seed varieties that extend the temporary hegemony of industrial agriculture – innovative bioengineering technologies with ever more unanticipated downsides (CRP).

Well fed: "Our world of global consumer culture is great fun – widescreen TVs, internet, convenient cell phones, global jet travel, the comfy fire retardant cushions of modern furniture, and heated chlorine-laced swimming pools of our ever expanding consumer culture. We are well fed, if, in fact, even a bit over-nourished, living within the growing safety net of our state security apparatus and our sophisticated electronic surveillance networks." (TH).

News Bites

100 billion gallons: "According to a recent study, conducted by the Environmental Working Group (EWG), just one drilling site [for natural gas] deploys harmful chemicals sufficient 'to contaminate more than 100 billion gallons of drinking water to unsafe levels… more than 10 times as much water as the entire state of New York used in a single day." (http://splashdownpa.blogspot.com/2009/09/is-hydrofracking-for-natural-gas-worth.html).

Abandoned boats: Folly Beach, Florida police have issued a warning that any boats found abandoned in the creeks and marshes around the city will be subject to salvage and recycling (Associated Press 2008). As the economy worsens, abandonment of boats is a growing national problem. "In Lee County Florida, deputies report hundreds, if not thousands, of abandoned boats clogging Southwest Florida waterways." (PR Web 2007).

After America optimism: "Starobin presents his farsighted and fascinating predictions for the After America world. These possibilities include a global chaos that could be dark or happy, a multipolar order of nation-states, a global Chinese imperium, or–even more radically—an age of global city-states or a universal civilization leading to world government." (Starobin 2009).

Alphabet soup: "CDOs, CDO squardes, CDSs, ABXs, CMBXs, are all components of the alphabet soup of synthetic financial instruments." (Soros 2008b).

Anaerobic sewage treatment: The flame retardant DECA PBDE (polybrominated diphenyl ether) widely used in computers and televisions and considered relatively inert, can be broken down by microbes such as anaerobic bacteria and dehalogenated into smaller highly toxic PBDE cogeners now classified as highly toxic persistent organic pollutants. "The levels of these compounds have been rising throughout the world, especially in North America, and their neurotoxic effects are similar to those of PCBs, which they resemble chemically." (Betts 2006).

Aquaculture in crisis: "Impacts include the introduction of non-native farmed fish species that diminish or replace indigenous fish populations; the propagation of deadly fish diseases; and the over-fishing of vast quantities of non-commercial fish to feed carnivorous farmed fish, such as salmon… Farmed fish often receive large doses of antibiotics to protect them from disease and are exposed to a variety of pesticides used to kill parasites and body fungi–all of which accumulate in the fish's tissues." (http://truefoodnow.org/campaigns/fish/).

Atmospheric transport of pollutants: "The power of the lidar remote sensing technique is best illustrated by considering that during the mission from Wallops Island, Virginia, to Bermuda in August 1982, as reported here, we measured the equivalent of $\sim 10^3$ individual profiles of aerosol with a vertical resolution of 15 m… these preliminary measurements on the vertical distribution of O_3, CO and aerosols over the western North Atlantic provide the first direct observation of how pollutants are being transported in both the boundary layer and free troposphere from source regions in the eastern United States to the North Atlantic." (Harris 1984).

Autism: "Autism affects 1 in 110 children, 1 in 70 boys." (http://www.autismspeaks.org/).

Bacteria subsisting on antibiotics: "Several antibiotics are natural products of microorganisms that have as yet poorly appreciated ecological roles in the wider environment... Bacteria subsisting on antibiotics are surprisingly phylogenetically diverse, and many are closely related to human pathogens... each antibiotic-consuming isolate was resistant to multiple antibiotics at clinically relevant concentrations. This phenomenon suggests that this unappreciated reservoir of antibiotic-resistance determinants can contribute to the increasing levels of multiple antibiotic resistance in pathogenic bacteria." (Dantas 2008).

Bats: The northeastern bat death toll hits 90%. Only 10% of the New England cave-dwelling bats have survived the massive die-off associated with powdery white fungal infection according to New York wildlife experts. (Baird 2009).

Biodiversity loss: Three decades ago there were thousands of seed companies and public breeding institutions worldwide. Today, 10 companies control more than two-thirds of global proprietary seed sales and almost 90% of agrochemical sales worldwide (http://www.etcgroup.org).

Biodiversity threats: The main threats to biodiversity reported in the pan-European region are climate change, urbanization and infrastructure development, agricultural intensification, land abandonment, desertification, acidification, eutrophication, radioactive contamination, forest fires, illegal logging, illegal hunting and wildlife trading, and invasive alien species (GEO 2007; Table 6, 8. pg. 237). Not mentioned by the UN publication editors is the broad impact of ecotoxin contaminant pulses from chemical fallout, which includes their ingestion and cross placental transfer in humans. Also not mentioned are the impact of genetically modified organisms, the loss of oceanic biodiversity, and mass extinction events such as bee colony collapse.

Biofixation of CO_2: Multipurpose processes using microalgae mass cultures can combine the direct capture of fossil CO_2 with the production of renewable fuels. Biofixation of CO2 can provide additional environmental services such as wastewater treatment, reclaiming water for agriculture, or the mass production of co-products such as biopolymers, fertilizers, and animal feed. All of these activities can significantly mitigate GHG (greenhouse gas) emissions and are particularly useful in developing countries (Biofixation Network; http://www.co2captureandstorage.info/networks/Biofixation.htm).

Biomonitoring: "The CDC spent $13.8 billion in its biomonitoring program in 2009." (Environmental Working Group 2009, 43).

Biopharmaceuticals: "Over time, products have increased in complexity, and the range of expression systems has broadened. The days of the blockbuster product isn't over, but much of the 'low hanging fruit' has been taken – so most future products are likely to reach smaller markets, as indicated by the size of the sales figure bubbles." (Scott 2009).

Bioplastics: "'Bioplastics' are often made from plant matter such as corn starch, potato starch, cane sugar, and soy protein… [and] have the long term benefits of reducing global warming pollution and our dependence on fossil fuels… Numerous concerns have been raised about their true value… The plant based material will actually contaminate the recycling process if not separated from conventional plastics… Some bioplastics cannot be broken down by the bacteria in our backyards… Plants grown for bioplastics have negative impacts of their own. Bioplastics are often produced from genetically modified food crops such as corn and soybeans, a practice that carries a high risk of contaminating our food supply [and are produced] using large amounts of chemical pesticides and fertilizers that pollute our air and water… The growth of the bioplastics and biofuels industry… rely on food crops as their raw material [and thus have a worldwide impact on agriculture.] Environmental advocates are calling for bioplastic production based on renewable crops (such as wild grasses) grown without chemicals, [as well as from] agriculture waste." (Union of Concerned Scientists 2009).

Bird population decline: "Just as they were when Rachel Carson published 'Silent Spring' nearly 50 years ago, birds today are a bellwether of the health of land, water and ecosystems," according to US Interior Secretary Ken Salazar. "From shorebirds in New England to warblers in Michigan to songbirds in Hawaii, we are seeing disturbing downward population trends that should set off environmental alarm bells." The species in decline are being affected by climate change, habitat destruction, invasive species and disease, among other factors (Eilperin 2009).

Bisphenol-A: A recent study by the Environmental Working Group (EWG) found Bisphenol-A (aka BPA) in more than half of 97 cans of brand name fruit, vegetables, soda, and other canned goods (*Body of Evidence*; www.ewg.org). "One of every three cans (33%) of infant formula contained BPA at levels where a single serving would expose an infant to more than 200 times the government's traditional safe level of exposure for industrial chemicals. The chemical is leaching from epoxy resins used to line cans." (http://newsgroups.derkeiler.com/Archive/Alt/alt.politics/2008-02/msg03534.html). "BPA is one of the chemicals with the highest volumes of production worldwide; global BPA capacity in 2003 was 2,214,000 metric tonnes." (Schuiling 2005, 42).

Bisphenol-A and 4-tertiary-octylphenol: "BPA was detected in 92.6% of persons ≥6 years of age [in the US] with total concentrations ranging from 0.4 µg/L to 149 µg/L. tOP was detected in 57.4% of samples at total concentrations of 0.2 µg/L-20.6 µg/L." (Calafat 2008).

Bisphenol-A ban: "MERI backs Bisphenol-A ban. 'With more than 3,000 miles of coastline and the largest lobster industry of any state, Maine needs to be aware that the impacts of BPA and related alkylphenol chemicals on the marine environment could be devastating'… After extensive, multiyear research, Hans Laufer of the University of Connecticut, has recently reported that similar compounds, known as alkylphenols, are possible culprits in the onset of crippling lobster shell disease." (*Mount Desert Islander* August 16, 2010, 9).

Black carbon: "Black carbon is the soot produced by billions of cooking fires and household stoves in the developing world. Scientists estimate it contributes half as much to warming the planet as the total carbon dioxide emissions from fossil fuels." (Cribb 2010, 127).

Body burdens: Laboratory tests sponsored by the Environmental Working Group "uncovered 171 pollutants in blood and urine from 9 volunteers, including an average of 56 carcinogens in each person." (www.ewg.org). Human exposure comes not only from the buildup of environmental contaminants in the food chain and from exposure to contaminants in air and drinking water, but also from the direct ingestion of substances used in food packaging and in processed foods, and from absorption of certain substances used in cosmetics through the skin (www.oztoxics.org/cmwg/body%20burden/load.html).

Breast cancer risk: "The risk that a 50-year-old white woman will develop breast cancer has soared to 12 percent today from 1 percent in 1975. (Some of that is probably a result of better detection.) Younger people also seem to be developing breast cancer." (Kristof 2009).

Breast milk: "From the early 70's, when they first appeared commercially, to 1998, levels of PBDE's in breast milk were doubling every five years, a rate unmatched by any known chemical in the last 25 years." (Williams 2005).

Breast milk biomonitoring: "In a new study of breast milk and store-bought milk from across the United States, scientists at Texas Tech University found perchlorate in every sample but one. The results suggest that this thyroid-disrupting chemical may be more widespread than previously believed." (American Chemical Society 2005).

Bubbles: "Bubbles always involve the expansion and contraction of credit and they tend to have catastrophic consequences." (Soros 2008b).

C. difficile infection (CDI - *Clostridium difficile*): "*C. difficile*, an anaerobic, gram-positive rod, is the most frequently identified cause of antibiotic-associated diarrhea (AAC). It accounts for approximately 15-25% of all episodes of AAC." (CDC; www.redorbit.com/modules/news/).

C. difficile infection rate: "The growth of infections caused by *Clostridium difficile* (c.dif) is an alarming trend in hospitals today. In 1993, an estimated 86,000 patients were discharged from hospitals with c. dif infections. By 2005, there were more than 300,000 CDI hospital discharges. Death from CDI has doubled in the past few years as one strain of *Clostridium difficile* has developed antibiotic resistance and evolved greater toxicity. A large-scale national prevalence study in 2008 revealed the CDI rate was 6.5 to 20 times greater than previous estimates." (Consumers Union 2008).

C. difficile update: "Several years ago, the mortality rate from a *C. difficile* infection was around 1 to 2 percent. But today, various studies estimate that the death rate is 6 percent. The reason is that a hypervirulent strain has emerged that emits higher levels of toxins than earlier strains." (Parker-Pope 2009).

CDC biomonitoring: In a survey of 8,373 individuals, methylmercury was found in 7,356 participants. In their survey of 694 individuals, 670 carried detectible levels of organochlorine (OCs) pesticides (CDC 2008).

CO_2 time delay: "Because of the long time scale required for removal of CO_2 from the atmosphere as well as the time delays characteristic of physical responses of the climate system, global mean temperatures are expected to increase by several tenths of a degree for at least the next 20 years even if CO_2 emissions were immediately cut to zero." (Friedlingstein 2005, 10832).

Calories: There are 65 calories of fossil fuel in every calorie of hamburger (TH).

Canadian officials contaminated: "The testing of Ontario Premier Dalton McGinty, progressive conservative leader John Tory, and NDP leader Howard Hampton revealed high levels of toxic chemicals from sources such as consumer products and industrial processes." (Environmental Defence, Canada. September 22, 2008; www.environmentaldefence.ca).

Cancer: WHO (World Health Organization) expects an annual increase of 1 percent in cancer cases and deaths, with higher rates in China, Russia, and India, leading to a projection of 27 million annual diagnoses and 17 million deaths by 2030. "…cancer already kills more people in poor countries than HIV, malaria and tuberculosis combined." (Douglas Blayney, MD, American Society of Clinical Oncology).

Cancer and epigenetics: "The evidence linking epigenetic processes with cancer is becoming 'extremely compelling' says Peter Jones, Director of the University of Southern California Norris Comprehensive Cancer Center." (Weinhold 2006).

Carbon Nanotubes: Carbon nanotubes "contain large percentages of carbonaceous and metal impurities… the metals and the catalytic materials used to synthesize the nanotubes may react with halogenated contaminants to generate radical species and toxic degradation products." (ES&T on line news April, 2008). Nanotubes grown in heated quartz reactor tubes produce toxic gases such as PAHs (polycyclic aromatic hydrocarbons), some of which escape collection by polyurethane foam (PUF) vessels.

Cats and chemical fallout: "The growing use of PBDEs in consumer products of the past 30 years has paralleled the rising incidence of feline hyperthyroidism, and a preliminary study suggests that PBDEs are found at higher levels in cats stricken with this disease (Mueller 2007). In addition to PBDEs, hyperthyroidism in cats could be linked to the plastics chemical and potent endocrine disruptor BPA that is known to leach from the pop-top cat food can lining into food." (Naidenko 2009).

Chemical body burden: The worldwide annual production of organic (carbon-based) chemicals has increased from about 1 million tons in the 1930s to 7 million tons in 1950, 63 million tons in 1970, and 250 million tons in 1985. It has been estimated that the annual production of organic chemicals will double every 7-8 years. This projection suggests an annual world production of 2 billion tons of organic chemicals by 2010. Since 1945, both the volume of chemicals used and the number of chemicals have increased dramatically, with around 80,000 chemicals currently in widespread use. This rapid post-war expansion in the chemical industry has certainly resulted in increased fetal exposure to many lipophilic contaminants (http://www.oztoxics.org/cmwg/body%20burden/load.html).

Children: "Childhood rates of chronic health problems, including obesity, asthma and learning disabilities, have doubled in just 12 years, a new study reports – to 1 in 4 children in 2006, up from 1 in 8 in 1994… Many chronic health problems resolve themselves during childhood. While half of the children followed from 2000 through 2006 had a chronic condition at some point during the period, only one-quarter did at the study's end." (Rabin 2010).

Chlorinated solvents: "Along with trihalomethanes, the VOCs (volatile organic compounds) methylene chloride, PCE, TCA, and TCE are the most frequently detected chlorinated solvents in the nation's aquifers." (USGS 2006).

Cholera: The Central Hospital in Zimbabwe's capital – Harare – officially closed down and now hardly a doctor or nurse is in sight, Zimbabwean journalist Brian Hungwe reported. Cholera-sufferers would be "coming to hospital to die because there is nobody to care for anyone," said Dr Malvern Nyamutora, vice-chairman of the Junior Doctors' Association.

Club of Rome paradigm: "In this run the collapse occurs because of nonrenewable resource depletion. The industrial capital stock grows to a level that requires an enormous input of resources. In the very process of that growth it depletes a large fraction of the resource reserves available. As resource prices rise and mines are depleted, more and more capital must be used for obtaining resources, leaving less to be invested for future growth. Finally investment cannot keep up with depreciation, and the industrial base collapses, taking with it the service and agricultural systems, which have become dependent on industrial inputs (such as fertilizers, pesticides, hospital laboratories, computers, and especially energy for mechanization)." (Meadows 1972).

Coal ash: "The coal ash pond that ruptured and sent a billion gallons of toxic sludge across 300 acres of East Tennessee last month was only one of more than 1,300 similar dumps across the United States – most of them unregulated and unmonitored – that contain billions more gallons of fly ash and other byproducts of burning coal. Like the one in Tennessee, most of these dumps, which reach up to 1,500 acres, contain heavy metals like arsenic, lead, mercury and selenium, which are considered by the Environmental Protection Agency to be a threat to water supplies and human health. Yet they are not subject to any federal regulation, which experts say could have prevented the spill, and there is little monitoring of their effects on the surrounding environment." (Dewan 2009).

Colony collapse disorder (CCD): "Recently approved insecticides have been implicated in recent massive bee colony die-offs and are the focus of increasing scientific scrutiny... The EPA has failed to respond to NRDC's Freedom of Information Act request for agency records concerning the toxicity of pesticides to bees." (Organic Consumers Association 2009; www.organicconsumers.org/articles/article_14192.cfm).

Comatose: "The state of the Union is comatose… the gridlock isn't only a function of paralyzed politics and special interests. There's been a gaping leadership deficit… both parties are dysfunctional … GOP's record of fiscal recklessness and its maddog obstructionism (are among)… the calamities left behind." (Rich 2010).

Contaminants in Maine birds: "We found both established (Hg, PCBs, chlordane, HCB, DEE) and emerging (PBDEs, PFCs) bioaccumulative toxic pollutants of concern in all the bird eggs we analyzed. Our results are the first records of PFCs in Maine birds." (Goodale 2008, 35-6). See *Appendix M* in *Volume 3*.

COREXIT 9500: "Data published by Environment Canada, that country's main environmental agency, showed common household dish soap as having a substantially higher rainbow trout toxicity than COREXIT 9500. Put another way, COREXIT 9500 is the more than 27 times safer than dish soap." (Nalco; http://www.nalco.com/news-and-events/4255.htm).

Corporate lobbying in political elections: "Justices, 5 – 4, reject corporate campaign spending limit. Dissenters argue that ruling will corrupt democracy." (Headline, January 22, 2010, *The New York Times*).

Default insurance: "Banks bet Greece defaults on debt they helped to hide... A new form of trading complicates efforts to borrow money. Bets by some of the same banks that helped Greece shroud its mounting debts may actually now be pushing the nation closer to the brink of financial ruin." (Schwartz 2010, A1).

Dengue Fever: "New Delhi – An epidemic of dengue fever in India is fostering a growing sense of alarm even as government officials here have publicly refused to acknowledge the scope of a problem that experts say is threatening hundreds of millions of people, not just in India but around the world." (Harris 2012, A8).

Diclofenac: Diclofenac is a potent anti-inflammatory drug given to cattle. Vulture death rates due to diclofenac poisoning on the Indian subcontinent are estimated to be as high as 99% and the species may soon be extinct. Feral dogs are now the main scavengers of ungulate carcasses, promoting the migration of diclofenac into the food web via the microbial digestion of contaminated scat. Filling the ecological niche of Asian vultures are an ever increasing numbers of Asian dog and rat populations, presenting new problems for human populations (rabies, lice, flea-borne illnesses) (*Wall Street Journal* May 1, 2008).

DINCH: "A new federal law took effect [in England] this week banning phthalates from children's toys and other kid's products. While the ban was hailed as a victory for children's health, it's no guarantee the products are safe. That's because companies aren't currently required to disclose the chemicals they use in place of phthalates... The German giant BASF started selling the plasticizer DINCH in 2002, and a company official says it is the most widely used phthalate substitute in the world. However, there are no peer-reviewed, publicly available data on the toxicity of DINCH, and what is widely known comes from animal studies conducted by the manufacturer and given to European food regulators. In those studies, BASF tested the chemical on rats and rabbits, and the results suggested that DINCH does seem to pose some risks to kidney health in animals." (Varney 2009).

Dioxin in fish oil: "Fishmeal and fish oil are the most heavily contaminated ingredients in feed production, with products of European fish stocks more heavily contaminated than those from South Pacific stocks. The fish stocks of concern in the north European industry are sprat and herring from the Baltic Sea, and herring, sprat, sand eel. and blue whiting in the North Sea. The differences in dioxin and PCB levels reflect the general

78

pollution level in the respective fishing areas and will disfavour the North European fishmeal and oil producers in the world market." (The Fishery and Aquaculture Industry Research Fund 2009; http://www.fiskerifond.no/index.php?current_page=prosjekter&subpage=archive&detail=1&id=611&gid=3).

Dispersant toxicity: "Available data indicate that Corexit 9500, Corexit 9527, and Corexit 9580 have moderate toxicity to early life stages of fish, crustaceans, and mollusks (LC50 or EC50 = 1.6 to 100 ppm)." (George-Ares 2010, 1007).

Domestic productivity: "The debt-driven financial services sector of the [US] economy grew by 2004 to constitute 21 percent of the US gross domestic product versus just 13 percent for manufacturing." (Phillips 2006).

Drinking water contamination: "Atrazine could be getting into water through food and drink, the researchers suggest, with, for example, many soft drinks containing corn syrup helping the pesticide to spread through the water-treatment system. However, like the other contaminants found by the team, the levels were below the US Environmental Protection Agency's safe maximum. For atrazine, this is 3.0 micrograms per litre; the highest value recorded by the researches was 930 nanograms (0.93 micrograms) per litre." (Lubick 2008).

Drugs: "As pharmaceutical use soars, drugs taint water and wildlife. With nearly $800 billion in drugs sold worldwide, pharmaceuticals are increasingly being released into the environment. The 'green pharmacy' movement seeks to reduce the ecological impact of these drugs, which have caused mass bird die-offs and spawned antibiotic-resistant pathogens." (Shah 2010).

Electronic waste: In a recent program on electronic waste being exported from America to China, CBS reported on the hazards of precious and heavy metal extraction from computer CPUs (central processing units), monitors, TVs, cell phones, and other electronic equipment. The CBS news story failed to include the highly mobile ecotoxins such as potent POPs (persistent organic pollutants) and EDCs (endocrine disrupting chemicals) noted in the landmark study, *Exporting harm: The high tech trashing of Asia* (Puckett 2002), which also accompany heavy metals in electronic waste. *Exporting Harm* also notes chlorinated dioxins, PVC (polyvinyl chloride), and brominated fire retardants (PBDE, TBBPA, PBB) as electronic waste ecotoxins of major concern (GP).

Electronic waste recycling: "Some 53 million tons of electronic waste was generated worldwide in 2009, according to ABI Research, a technology market research firm. Only about 13 percent of it was recycled." (Zeller 2010).

Endocrine disrupters: "While no clear links are yet evident, some believe increasing exposure to estrogen-like chemical mixtures in everyday life contributes to rising rates of testicular and breast cancers, increasing numbers of sex organ deformities, declining sperm counts, and precocious puberty in industrialized countries." (http://e.hormone.tulane.edu/learning/estrogens.html#estrogen_disrupters; April 15, 2010).

Europe: Europe crisis slams Ireland: Dublin to spend billions more to shore up lenders; jitters over Euro's future. (*The Wall Street Journal* October 1, 2010).

EWG/Commonweal Study #1: In this survey of the blood and urine of adults, 171 of 214 industrial ecotoxins were found. High levels of chlorinated dioxins and furans were noted in 78% of the participants; the other 22% of the participants had only moderate levels of these carcinogenic, teratogenic, and mutagenic ecotoxins. One hundred percent of the participants had moderate levels of PCBs as well as low levels of organochlorine pesticides. One hundred percent of adults tested also had un-quantified levels of phthalates and volatile organic compounds (VOCs) and semi-volatile organic compounds (SVOCs) (www.ewg.org).

Farmed Salmon: Farmed salmon carry up to ten times more cancer-causing chemicals than their wild cousins. The average dioxin level in farmed salmon was 1.88 parts per billion (ppb) vs. 0.17 ppb for wild salmon. The main source of PCBs in farmed salmon is the fish oil that constitutes a large part of their feed. Small forage fish such as herring are processed into salmon farm chow. Salmon on US farms are also fed recycled fat from slaughtered livestock, including cows. The average PCB level was 36.6 ppb in farmed salmon vs. 4.75 in wild salmon (Hopkin 2004).

Federal Deposit Insurance Corporation: "At F.D.I.C., bracing for a wave of failures." (*The New York Times* headline, February 24, 2010).

Financial globalization: "Financial globalization has definitely turned out to be even more dangerous than we realized." (Krugman 2008).

Fishermen's felt sole boots: "Growing scientific evidence suggests that felt, which helps anglers stay upright on slick rocks, is also a vehicle for noxious microorganisms [such as Didymosphenia geminata or didymo] that hitchhike to new places and disrupt freshwater ecosystems." (Barringer 2010).

Flat-world technology: The production of one silicon chip requires the use of 2,600 gallons of water (TH).

Foiling the Clean Water Act: "Thousands of the nation's largest water polluters are outside the Clean Water Act's reach because the Supreme Court has left uncertain which waterways are protected by that law, according to interviews with regulators…

The court rulings causing these problems focused on language in the Clean Water Act that limited it to 'the discharge of pollutants into the navigable waters' of the United States. For decades, 'navigable waters' was broadly interpreted by regulators to include many large wetlands and streams that connected to major rivers. But the two decisions suggested that waterways that are entirely within one state, creeks that sometimes go dry, and lakes unconnected to large water systems may not be 'navigable waters' and are therefore not covered by the act – even though pollution from such waterways can make its way into sources of drinking water… Companies that have spilled oil, carcinogens and dangerous bacteria into lakes, rivers and other waters are not being prosecuted, according to Environmental Protection Agency regulators working on those cases, who estimate that more than 1,500 major pollution investigations have been discontinued or shelved in the last four years." (Duhigg 2010).

Freon 13 ($CClF_3$): Over a period of 100 years, one pound of $CClF_3$ (freon 13) has the same potential to aggravate global warming as 8,100 pounds of CO_2 (http://www.epa.gov/fedrgstr/EPA-AIR/1995/October/Day-11/pr-1117.html).

GE contaminants: "Three rice contamination events this past year are only the latest example of farmers' incurring large financial losses due to the presence of GE contaminants." (Union of Concerned Scientists 2008, 15).

GE (genetically engineered) food and feed crops: "The two primary GE food and feed crops are corn and soybeans. GE soybeans are now grown on over 90 percent of soybean acres, and GE corn makes up about 63 percent of the U.S. corn crop." GE crops include corn containing transgenes from Bt (*Bacillus thuringiensis*) bacteria that confer resistance to several kinds of insects, corn containing transgenes for herbicide tolerance, and soybeans that contain a transgene for herbicide tolerance. "GE has done little to increase overall crop yields." (Gurian-Sherman 2009b, 1).

General alarm fire: Halogenated (brominated) flame retardants produce highly toxic dioxin-furans when incinerated, as is frequently the case when discarded electronic equipment is burned in Asian city dumps after valuable heavy metals have been salvaged from discarded computers and other electronic equipment (JW).

Genetically engineered salmon: "Research published in the *Proceedings of the National Academy of Sciences* notes that a release of just sixty GE fish into a wild population of 60,000 could lead to the extinction of the wild population in less than 40 fish generations. Wild Atlantic salmon are already on the Endangered Species List in the U.S.; approving these GE Atlantic salmon will be the final blow to these wild stocks." (Center for Food Safety; http://ge-fish.org/).

Glacial earthquakes: Glacial earthquakes in Greenland are the probable predicator of the breakup of the Greenland ice sheet and its catastrophic slippage into the sea. The possible result: a sudden 20' increase in world sea levels with a massive North Atlantic tsunami (CRP).

Global arms trade: "Today, America has a quarter of a million troops and civilians stationed in 130 countries. It is, by far, possessor of the largest military establishment in the world and is the world's largest arms exporter. (The U.S. share of the global arms trade doubled after the Cold War ended, so that America now sells roughly half of all the weapons sold worldwide." (Berman 2006, 143).

Global warming potential: "Methane has 23 times the global warming potential (GWP) of CO_2 and nitrous oxide has 296 times the warming potential of carbon dioxide." (Knickerbocker 2007).

GM corn: An Austrian study on mice has shown that genetically modified (GM) corn has a damaging effect on the reproductive system. Over 20 weeks, with an approved GM product, it became clear that the fertility of mice fed GM corn was seriously impaired, with fewer offspring than mice fed non-GM, equivalent material. Mice fed GM maize had a statistically significant lower number of offspring and a decrease in litter weight in the third and fourth generations. Mice fed GM-free corn reproduced more rapidly (http://www.greenpeace.org/international/press/releases/ge-threat-to-fertility-11112008).

Good news on lead: "The removal of lead from gasoline resulted in an 80 percent decline in lead levels in our blood since 1976 – along with a six-point gain in children's I.Q.'s, Dr. Landrigan [Mt. Sinai] said." (Kristof 2009).

Gordon Brown: "As we have discovered to our cost, the problem of unbridled free markets in an unsupervised marketplace is that they can reduce all relationships to transactions, all motivations to self-interest, all sense of value to consumer choice and all sense of worth to a price tag." (*The New York Times* March 29, 2009).

Government debt: "The White House estimates government debt accounted for 90 percent of the economy's total output in 2009, up from 70 percent a year earlier. While the cost of borrowing for the government and others remains historically low today, it could surge higher in the coming years." (Bajaj 2009).

Greenhouse gasses: "The world currently emits about 55 billion U.S. tons of carbon dioxide equivalent per year, adding about 1 ppm of greenhouse gasses to the atmosphere every four months. At the present rate it will reach 450 ppm around 2030." (Cribb 2010, 144).

Guano: "The major source of phosphorus fertilizers is from the mining of guano. Current estimates project readily accessible sources of guano as being depleted by the end of the 21st century." (McGraw-Hill 2002, 72).

Gunslinging trading operators: "By the early 2000s, the hedge fund industry was poised for a phenomenal run that would radically change the investment landscape around the world. Pension funds were diving in, and investment banks were expanding their proprietary trading operations… Hundreds of billions poured into the gun slinging trading operations that benefitted from an age of easy money, globally connected markets on the Money Grid, and complex quantitative strategies." (Patterson 2010, 151). "Most banks, inspired by Salomon Brothers, had created off balance sheet vehicles that would temporarily house the loans as they were bundled, packaged, sliced, diced and sold around the world." (Patterson 2010, 202).

Habitat degradation: "Winter flounder is an important commercial and recreational fish throughout New England and the Mid-Atlantic. Inshore habitat degradation and overfishing have contributed to serious stock declines throughout the species' range, leaving both fisheries at a fraction of their historical numbers." (Berger 2009, 35).

Halliburton loophole: "Oil and gas companies that use hydraulic fracturing are exempt from regulations under the Safe Drinking Water Act that would require them to disclose the cocktail of chemicals they use." (http://www.americanrivers.org/newsroom/blog/hydrofracking-poses-threats-1-10.html).

Hazardous waste: "The estimated production of hazardous chemical-based pollutants in the US by industry alone is more than 36 billion kilogrammes/year, with about 90% of these chemicals not being disposed of in an environmentally responsible manner." (United Nations 2006, 135). It should be noted this is a conservative estimate that focuses only on direct industrial production of extremely hazardous chemicals and does not include emissions from consumer products and electronic wastes, numerous emerging chemicals of concern, pharmaceuticals, remobilized naturally occurring ecotoxins, such as methylmercury, military ecotoxin point sources of all kinds, and other point sources. A more accurate estimate of annual US ecotoxin production would be 200 billion kilograms or more than 200 million tons (JW).

HIV/AIDS: "About 42% of pregnant women in Swaziland are HIV-positive, an increase of 3% since last year… the increase likely is because more women's lives are being prolonged through improved access to antiretroviral drugs. About 185,000 people in Swaziland -- which has a population of one million and the highest HIV/AIDS prevalence rate worldwide -- are living with HIV. About 30,000 people have access to antiretrovirals in the country, and average life expectancy is 37 years (*Kaiser Daily HIV/AIDS Report* February 23, 2009).

Hormonal chaos: "New evidence suggests that relatively low levels of industrial chemicals may mimic or obstruct hormonal activity – with potentially devastating long term effects that range from cancer and reproductive abnormalities to cognitive dysfunctions like attention deficit disorder." (Krimsky 2000).

Household debt: "Total [US] household debt, which includes both consumer debt and mortgage debt, jumped from 6.5 trillion in 2000 to 10.2 trillion in 2004." (Phillips 2006, 328). By 2008, US household debt was approaching $50,000 for every US citizen, without any consideration of accumulated national debt or commercial debt (JW).

Human adipose tissue: A study by Onstot et. al in 1987 involving HRGC/MS analysis of human adipose tissue revealed approximately 700 likely toxic contaminants (Onstot 1987).

Infants at risk: "In February 2008, U.S. Food and Drug Administration said that U.S. toddlers on average are being exposed to more than half of the U. S. EPA's safe dose [of perchlorate] from food alone. In March 2009, a Centers for Disease Control study found 15 brands of infant formula contaminated with perchlorate. Combined with existing perchlorate drinking water contamination, infants could be at risk for exposure to perchlorate above the levels considered safe by E.P.A." (www.wikipedia.org 2010).

Lifting shadows: The lifting shadows of what transpires in the shadow banking network reveal a sophisticated world of instant electronically formulated synthetic collateralized debt obligations and hedge fund scams, the non-sustainable components of finance capitalism teetering on the edge of bankruptcy (CRP).

Marine mammal tissues TCP: "Concentrations of 4,4'-DDE were significantly higher in common murre eggs from St. Lazaria and East Amatuli islands in the Gulf of Alaska (2440 ± 800 and 1570 ± 740 ng/g lipid weight, respectively) than eggs from the Bering Sea colonies (Little Diomede, St. George, and Bogoslof islands) and in concentrations reported by Braune et al. (2001) from Prince Leopold Island thick-billed murre colonies (775 ± 54 ng/g lipid weight)." (Alaska Marine Mammal Tissue Archival Project; http://www.absc.usgs.gov/research/ammtap/ammtap.htm).

Marine mammals: "The transfer of anthropogenic contaminants within the marine ecosystem occurs between the aqueous matrix and the living biota and the biotic components of the food chain, with the food-based transfer dominating the higher trophic levels. Arctic food webs are relatively simple, and within the context of the existing data, offer a natural laboratory to test developing hypotheses about predator-prey patterns of bioaccumulation."
(http://www.absc.usgs.gov/research/ammtap/summary.htm).

Mass extinction events: "According to University of Chicago paleontologist, David Raup, the background rate of extinction on Earth throughout biological history has been one species lost every four years on average. According to one recent calculation, human caused extinction may now be running as much as 120,000 times that level." (30,000 per year) (Bryson 2003).

Medical bankruptcy: "Between 1981 and 2001, medical bankruptcies increased by 2,200%." (Phillips 2006, 294).

Memory chip energy: "The embodied energy of the memory chip alone already exceeds the energy consumption of a laptop during its life expectancy of 3 years." (De Decker 2009).

Mercury: "…a years-long stalemate on mercury emissions appears to have broken after the White House Council on Environmental Quality issued a statement saying, 'The United States will play a leading role in working with other nations to craft a global, legally binding agreement that will prevent the spread of mercury into the environment.'" (Pope February 23, 2009 *Sierra Club Insider*).

Mercury from the atmosphere: "The Casco Bay atmospheric deposition studies indicate that the atmosphere is the major contributor of mercury and the likely source of 30% or more of the PAHs [polycyclic aromatic hydrocarbons] that enter the coastal ecosystem. Pollutants can be deposited from nearby sources or can travel from other regions of the country via wind and precipitation." (Casco Bay Estuary Partnership 2008).

Mercury/PCB Synergism: "… evidence is building that mercury and PCBs together are much more toxic than either one individually or simply added together. Some of the learning disabilities seen in Great Lakes children studies may be due to the multiplied health effect (synergism) of these chemicals together." (http://www.foxriverwatch.com/dioxins_pcb_pcbs_1.html).

Mercury in women's blood: "The level of inorganic mercury in the blood of American women has been increasing since 1999 and it is now found in the blood of one in three women, according to a new analysis of government data for more than 6,000 American women… As a result of chronic mercury exposure, between 300,000 and 600,000 American children were born with elevated risks of neurodevelopmental disorders between 1999 and 2000." (Environment News Service 2009).

Methicillin-resistant staphylococcus aureus (MRSA): Australia is facing an epidemic of drug-resistant staphylococcal infections, also known as MRSA superbugs that attack healthy teenagers and can be fatal. Of particular concern is a new virulent form that can lead to a severe form of pneumonia and cause death in up to 50% of cases. The community strain of the bacteria carries more highly toxic genes than "hospital-acquired" MRSA. Outbreaks in the U.S. have led to the deaths of several teenagers (Stark 2008).

Methylmercury: "It is estimated that over 410,000 children born each year in the United States are exposed in the womb to methylmercury (MeHg) levels that are associated with impaired neurological development (Mahaffey 2005). Eight percent of US women of childbearing age have blood Hg levels in excess of values deemed safe by the US Environmental Protection Agency. (USEPA-Schober, et al. 2003)." (Driscoll 2007 *Bioscience*).

Methylmercury: "Studies on cod and salmon show that a large portion of the mercury in fish feed accumulates in the edible filet of farmed fish." (http://www.foodandwaterwatch.org/fish/seafood/chemicals-of-concern/).

Modified corn study: "Rats fed either genetically engineered corn or the herbicide Roundup had an increased risk of developing tumors, suffering organ damage and dying prematurely." (Pollack 2012, B5).

Moscow RADs: "According to Moscow City Hall figures, 11 nuclear research reactors currently operate in the city. More than 2000 organisations are using about 150,000 sources of ionising radiation, and almost 90 percent of them have exceeded their predicted periods of service." (*Radioactive waste threatens Moscow*; www.bellona.org).

MRSA in pork chops: Canadian researchers have found methicillin-resistant *Staphylococcus aureus* (MRSA) bacteria in just under 10% of sampled pork chops and ground pork products purchased in retail stores across the country. A group of Dutch researchers reported last fall that they had isolated MRSA from two pork samples in the Netherlands. And Japanese scientists reported in 2005 that they had found MRSA in two samples of raw chicken (Branswell 2008).

MRSA rates: MRSA infection rates are approaching 1% of hospital admissions.

Myxomatosis: Myxomatosis is a naturally occurring virus in rabbits, which was intentionally introduced in Australia in 1950 in an effort to control the rabbit population. It has now spread worldwide decimating much of the world's rabbit and natural rabbit predator populations (JW).

Nanosilver: "A 2008 nanosilver legal petition to the United States Environmental Protection Agency (USEPA) filed by the International Center for Technology Assessment in Washington, DC identified more than 260 nanosilver products currently on the market, including household appliances and cleaners, clothing, cutlery, children's toys and personal care products." (Friends of the Earth 2009, 5).

National lake fish tissue study: Forty nine percent of EPA samples of fish from 36,000 lakes had mercury tissue concentrations that exceeded the 0.3 ppm screening value for mercury; 17% of the samples from 13,000 lakes had PCB concentrations that exceeded the 12 ppb screening value for total PCBs. Eight percent of the sample population of

6,000 lakes had dioxin and furan tissue concentrations that exceeded 0.15 ppt screening value. (Environmental Protection Agency 2009b).

Nepalese water sheds: Fully forested until 1950, most Nepalese watersheds were clear cut by 1990, rapidly depleting a major source of fresh water for many Asian urban areas (GP).

Neurotoxicants in human breast milk: "The detection of CDDs, hexachlorobenzene, *p,p'*-DDE, methylmercury, PCBs, and other potential neurotoxicants in samples of human breast milk has led to epidemiological studies of possible neurological deficits in children exposed *in utero* and during nursing to persistent chemicals in breast milk... Weight-of-evidence analyses of available data on the joint toxic action of mixtures of these components indicate that scientific evidence for greater-than-additive or less-than-additive interactions among these components is limited and inadequate to characterize the possible modes of joint action on most of the pertinent toxicity targets." (Agency for Toxic Substances and Disease Registry 2004, 101).

Nitrogen: "A cloudy future for Great Bay. Debate over wastewater, nitrogen rules continues. Nitrogen levels in the Great Bay Estuary have risen to a point that puts vital plant life at risk, and state officials and environmental experts say action needs to be taken. The primary solution under consideration, however, is expensive — many towns are facing wastewater plant upgrades that could cost up to $100 million." (Keefe 2010).

Obesity: "it is estimated that overall caloric intake in the U.S. has increased an alarming 600 calories per person per day since 1970. Burning these extra calories would require, on average, an additional hour of physical activity per day." (Landrigan 2010b).

Oil extraction peak: The current consensus among 18 recognized estimates of oil supply profiles is that the peak of extraction will occur in 2020 at the rate of 93 million barrels per day (mbpd). (www.energywatchgroup.org). "Pessimistic predictions of future oil production operate on the thesis that either the peak has already occurred, oil production is on the cusp of the peak, or that it will occur shortly." (http://en.wikipedia.org/wiki/Peak_oil#cite_note-deffeyes012007-4).

Organic farming: "UNEP [United Nations Environmental Programme] reported that organic practices in Africa outperformed chemical-intensive conventional farming, and also provided environmental benefits such as improved soil fertility, better retention of water, and resistance to drought. This analysis of 114 farming projects in 24 African counties found that organic or near organic practices resulted in a yield increase of more than 100%." (www.rodaleinstitute.org; citing UNEP 2008 http://www.unep.ch/etb/publications/insideCBTF_OA_2008.pdf).

Peak of the boom: "Rising prices also tend to generate optimism and encourage greater use of leverage-borrowing for investment purposes. At the peak of the boom, both the value of the collateral and the degree of leverage reach a peak." (Soros 2008b).

Perchlorate: "In a related 2006 study, the CDC found perchlorate in the urine of every one of 2,820 people tested, suggesting that food is a key route of exposure in addition to drinking water." (http://www.ewg.org/node/20968).

Perchlorates in breast milk: "Perchlorate inhibits iodide uptake and may impair thyroid and neurodevelopment in infants. Recently, we unambiguously identified the presence of perchlorate in all seven brands of dairy milk randomly purchased from grocery stores in Lubbock, TX. How widespread is perchlorate in milk? Perchlorate in 47 dairy milk samples from 11 states and in 36 human milk samples from 18 states were measured. Iodide was also measured in a number of the samples. Perchlorate was detectable in 81 of 82 samples. The dairy and breast milk means were, respectively, 2.0 and 10.5 microg/L with the corresponding maximum values of 11 and 92 microg/L." (Kirk 2005).

Pharmaceutical contaminants in treated drinking water: In Philadelphia, 56 pharmaceuticals or byproducts were found, including medicines for pain, infection, high cholesterol, asthma, epilepsy, mental illness, and heart problems. Some drugs, including widely used cholesterol fighters, tranquilizers, and anti-epileptic medications, resist modern drinking water and wastewater treatment processes. The EPA says there are no sewage treatment systems specifically designed to remove pharmaceuticals (Donn 2008).

Pharmaceuticals: "We know we are being exposed to other people's drugs through our drinking water, and that can't be good," says Dr. David Carpenter, who directs the Institute for Health and the Environment of the State University of New York at Albany. Over the past five years, the number of U.S. prescriptions rose 12 percent to a record 3.7 billion, according to IMS Health and The Nielsen Co. (JW).

Phosphorous as a limiting factor: The depletion of phosphorous fertilizers derived from the guano deposits of the Galapagos Islands has the potential to be a major limiting factor for the future productivity of the industrial agricultural ecosystems of global consumer society. Guano deposits are expected to be completely eliminated by 2100 and no other major deposits for industrial agriculture have been located (GP).

Pilgrim Nuclear Power Station: "The spent fuel pool contains some of the largest inventories of radioactivity on earth. Pilgrim's pool has 3,759 assemblies in a space designed for 880. The pool would catch fire if its water drops below the top of the assemblies." (Paul Rifkin).

Polycyclic aromatic hydrocarbons (PAHs): "The PAHs in petroleum mixtures are of greatest concern for human health because of their persistence (i.e. lower evaporation rates), and their potential for toxic or carcinogenic effects." (US FDA 2010).

Powdered infant formula: "Researchers from the U.S. Centers for Disease Control and Prevention (CDC) have reported that 15 brands of powdered infant formula are contaminated with perchlorate, a rocket fuel component detected in drinking water in 28 states and territories. The two most contaminated brands, made from cow's milk, accounted for 87 percent of the U.S. powdered formula market in 2000, the scientists said… A series of studies have found perchlorate in the urine of every American tested by the CDC and in breast milk (Blount et al 2006b, Pearce et al 2007)… 2008 federal Food and Drug Administration (FDA) tests… found almost 75 percent of food and beverage samples tainted with perchlorate." (Jacob April 2009; http://www.ewg.org/node/27772/).

Preservation of Antibiotics for Medical Treatment Act (PAMTA): The PAMTA "would help combat the antibiotic resistance crisis America is currently facing. The bill would require the FDA to review prior approvals for antibiotics such as penicillin and tetracycline to determine whether they can be safely used as animal feed additives." (http://www.keepantibioticsworking.com/new/resources_library.cfm?RefID=104544).

Public wells: "An important source of chloroform and other THMs [trihalomethanes] in drinking-water supply wells may be the recycling of chlorinated water and wastewater. Mixtures of THMs commonly occur in public well samples, and the most frequently occurring are combinations of the brominated THMs." (Ivahnenko 2006).

Quagga mussel: An invasive species of mussel, probably originating from the Caspian Sea, which has now invaded the Great Lakes and is currently decimating the white fish population. This mussel disrupts the food chain by destroying plankton and interrupting the life cycle of a "shrimp-like organism known as *Diporeia*," which is essential for maintaining harvestable populations of the Lake Michigan white fish. It also results in the proliferation of large amounts of the slime-like algae *Cladophora*, now responsible for pervasive Great Lakes algae blooms threatening all levels of the Great Lakes food chain. (Quraishi 2011).

Readiness: A new report concludes that, as a result of budget cuts and the financial crisis, the US is not prepared for a rapid response in several critical areas including surge capacity, rapid disease detection, and food safety due to disease outbreaks, natural disasters, and bioterrorism. More than half of states and the District of Columbia scored seven or less out of 10 key indicators used to assess health emergency preparedness capabilities (Vinter 2008).

RTG (radioisotope thermoelectric generator): In 1964, a SNAP 9A RTG power source disintegrated 50 kilometers above the earth's surface and released 17,000 ci of Pu_{238}, tripling the worldwide inventory of this isotope and increasing the world's total environmental inventory (excluding weapons stockpiles or nuclear waste inventories) of all plutonium by 4% (CBM).

Safe Chemicals Act: "The Safe Chemicals Act was introduced by Senator Lautenberg (D-NJ) on April 15, 2010. The bill would amend the Toxic Substances Control Act (TSCA) to revise how the Environmental Protection Agency regulates chemicals. It has been referred to the Senate Committee on Environment and Public Works where it awaits further consideration." (Senator Susan Collins May 24, 2010).

Safety assessments: "Out of 12,500 different ingredients in cosmetics and personal care products, nearly 90% have not been assessed for safety by any accountable entity." (Environmental Health Strategy Center 2010, 1).

Safety testing: Major gaps in our system of public health protection allow most industrial chemicals on the market with no mandatory safety testing. "The U.S. Environmental Protection Agency (EPA), which oversees TSCA [Toxic Substances Control Act], has been able to require safety testing for just 200 chemicals in 34 years, and has only been able to restrict the use of 5 chemicals under the law." (Environmental Health Strategy Center 2010, 5).

Salmon virus: "The often fatal farmed Atlantic salmon (*Salmo salar L*) disease known as heart and skeletal muscle inflammation (HSMI) may be posing a risk to wild fish coming in close proximity to marine pens and escaped farmed fish. First detected in salmon on a farm in Norway in 1999, HSMI has now been found in 417 fish farms there as well as in the UK. The disease ravages heart and muscle tissue and kills up to 20 per cent of infected animals. Attempts to identify the pathogen causing the disease have been fruitless recently. But now cutting-edge molecular techniques have led researchers to new results. The disease may be caused by a previously unknown virus, according to an international team led by W Ian Lipkin, MD… This virus is related to previously known reoviruses, which infect various vertebrates." (Real 2010).

Salmonella: "FDA determined there was a public health risk in part because there had been a rapid rise in resistance to cephalosporin drugs in the food borne pathogen *Salmonella* in both humans and farm animals. Cephalosporin is the treatment of choice for serious *Salmonella* infections, as well as many other important infections, in humans that cause 1,300,000 U S illnesses each year. Rising resistance is a problem because it leads to more and more severe illness including more hospitalizations and more death." (http://www.keepantibioticsworking.com/new/resources_library.cfm?RefID=104544). "The extralabel use of cephalosporins in food animal medicine presents a clear risk to human health and should be restricted." (Margaret Mellon, Union of Concerned Scientists).

Science careers: "China is expected to surpass the United States in numbers of engineering doctorates by 2010. At the college level, statistics show a waning interest among US students in science-related careers; in 2001, only 17 percent of all bachelor degrees in the United States were in natural science and engineering, compared to a world average of 27 percent and a Chinese average of 52 percent." (Phillips 2006, 381).

Sea-level rise: "Regional sea surface temperatures have increased almost 2° Fahrenheit since 1970, and the rate of sea-level rise has intensified. Tide-gauge records in Portland, Maine, show a local relative sea-level rise of approximately eight inches since 1912." (Jacobson 2009).

Sentinel organisms: "Environmental research has suggested that some bivalves may be valuable as sentinel organisms for indicating levels of pollutants in coastal marine waters. These organisms concentrate numerous pollutants to a marked degree over sea-water levels. Several species thus 'bioaccumulate' some, or even most, members of the four identified categories of marine pollutants: heavy-metals, transuranic elements, petroleum hydrocarbons, and halogenated hydrocarbons. A given species, however, may have unique enrichment factors for any or all of these groups of substances." (Goldberg 1978, 101).

Sniper rifles: "In the long gun market, sniper rifles are the latest trend. Radically different from typical hunting firearms, sniper rifles boast breathtaking accuracy, range, and power. The weapon is capable of pinpoint shots from distances of up to 2,000 yards—or the length of 20 football fields. Even at such long range, the bullets can penetrate armor."
(http://www.consumerfed.org/elements/www.consumerfed.org/file/health/industry.pdf).

Stratospheric water vapor: Using balloon-borne frost point hygrometers, the GMD (Global Monitoring Division) of the NOAA (National Oceanography and Atmospheric Administration) has detected approximately a 1% per year increase in stratospheric water vapor at Boulder, CO since 1980 (www.esri/noaa.gov).

Superbugs: "Rates of [superbug] resistance to last-line antibiotics by a bacteria called *Klebsiella pneumoniae* had more than doubled to 15 percent by 2010 from around 7 percent five years ago. 'What's even more worrying is that there's a great diversity among different countries in Europe – and some countries have resistance of almost 50 percent.' Sprenger [also] said the report found that the countries with the highest rates of multi-drug resistant infections, such as Greece, Cyprus, Italy, Hungary and Bulgaria, also tended to be the ones with the highest use of antibiotics." (Kelland 2011).

Superbugs from Asia: "In a separate risk report on a gene known as New Delhi metallo-beta-lactamase, or NDM-1, that makes bacteria highly resistant to almost all drugs, the ECDC said 106 cases of infection involving the gene had been reported in 13 European countries by the end of March 2011. This is an increase on the 77 cases found in the same 13 countries in late 2010, with the new cases in Britain, France, Germany, Sweden, the Netherlands and Slovenia. The majority of the cases -- 68 out of the 106 -- were in Britain, the ECDC said, and 25 of those were in people who had travelled to or been hospitalized in India or Pakistan." (Kelland 2011).

Superweeds: "Superweeds immune to Roundup have spread to millions of acres in more than 20 states in the South and Midwest." (Kilman 2011).

Surveillance networks: "CDC recently established seven domestic and global sentinel surveillance networks linking health care providers to detect and monitor emerging diseases." (www.medicalnewstoday.com/printerfriendlynews.php?newsid=6971).

Teflon toxicosis: "Teflon toxicosis has been the cause of death for hundreds of pet birds nationwide whose lungs filled with blood after they breathed in toxic fumes from overheated, non-stick pans." (EWG 2003a; NRC 1991).

Thermal inertia: "The movement toward global warming – is now about 1.8 W/m^2 over the entire surface of the Earth. This means that every square meter – roughly the surface size of the desk I'm working on right now – of the planet is absorbing 1.8 watts more energy than it was in 1880." (Hartmann 2009, 29).

Top secret chemicals: "Under the 1976 Toxic Substances Control Act (TSCA), the chemical industry has been allowed to stamp a 'trade secret' claim on the identity of two-thirds of all chemicals introduced to the market in the last 27 years… These include substances used in numerous consumer and children's products… The public has no access to any information about approximately 17,000 of the more than 83,000 chemicals on the master inventory compiled by the EPA." (Andrews 2009, 2). "The total production of secret chemicals has also drastically increased. In 2006, EPA reported that the volume of secret chemicals produced or imported totaled between 1.265 billion pounds and 4.473 billion pounds, five to six times the 1990 volume of 255 million pounds to 995 million pounds." (Andrews 2009, 7).

Triclosan: "Studies indicate that in surface waters, triclosan can interact with sunlight and microbes to form methyl triclosan, a chemical that may bioaccumulate in wildlife and humans." (Adloffsson-Eresi 2002; Lindstrom 2002). "A recent European study found methyl triclosan in fish, especially concentrated in fatty tissue." (Balmer 2004). "Triclosan can also degrade into a form of dioxin." (www.ewg.org/sites/humantoxome/chemicals).

U. S. income: "Between 1978 and 2007, the share of U.S. income accruing to the top 1 percent of American families jumped from 9 to 23.5 percent of the total." (Fukuyama 2011, 23).

Ultrathin graphene: "conducts electricity better than copper, is up to 300 times stronger than steel and could be used to build better display screens… could also replace and redefine components in computers and phone by, for example, making them foldable." (*The New York Times*, January 28, 2013).

Ventilation: "Two-thirds of single-family homes built in California in recent years had substandard indoor air quality and excessive formaldehyde levels, partly because residents didn't open their windows for ventilation." (*USA Today* 2009).

Viral hemorrhagic septicemia (VHS): "Since 2005, the list of species known to be affected by [VHS] has risen to more than 40, including a number of ecologically and recreationally important fish." (USDA 2006, 1-2).

Wishful thinking at the Copenhagen Accord: "The United States agreed to reduce greenhouse gas emissions 17% from 2005 levels by 2020." (*The New York Times* January 1, 2010).

World debt: In 2007, total gross world productivity (GWP) was 47 trillion dollars and the total value of all stocks and bonds was 119 trillion dollars. In contrast, fueled by the activities of a shadow banking network and its financial engineers, total world debt in the form of highly leveraged derivatives, was 473 trillion dollars. By 2008, total world debt had grown to 596 trillion dollars (Ferguson 2009a).

Aphorisms and Epiphanies

Achilles heel: The Achilles heel of an ever-growing world population is the ongoing and accelerating contamination of the global atmospheric water cycle.

Additive joint toxic action: "There are several reasons supporting the recommendation to use component-based approaches that assume additive joint toxic action in exposure-based assessments of possible noncancer or cancer health hazards from oral exposure to mixtures of CDDs, hexachlorobenzene, *p,p*'-DDE, methylmercury, and PCBs." (Agency for Toxic Substances and Disease Registry 2004, 101).

Affordable oil supplies: "Affordable oil supplies are located in the least politically stable regions of the world." (Gore 2007).

Antibiotics: Regarding a ubiquitous antibiotic in hand soaps and cleaning agents, "the amount of Triclosan in the wastewater stream is estimated to be as much as 3 to 5 milligrams per day from residences alone." (McAvoy 2002).

Armageddon: Extinction by planetary collision is one of many end of the world apparitions; biocatastrophe is an ongoing round-world (not flat-world) scientific and historic event.

Assets: In the age of biocatastrophe, the increasing expense of toxic assets (real estate, derivatives, federal and credit market debt, chemical fallout residues, oil slicks, emerging antibiotic resistant viruses, etc.) is the counterpart of the decreasing availability of funding for the maintenance of public safety, health, cultural, educational, and historic assets, in the context of the rapidly dwindling natural resources of the World Commons.

Bad news: The rapid electronic transfer of bad news increases social stress.

Biocatastrophe: The symbiotic interrelationships between human and natural ecosystems culminate in biocatastrophe.

Biocidal silver: "Silver is comparatively more toxic than other heavy metals when in nanoparticle form." (Braydich-Stolle 2005). The greater bioavailability of nanoparticles, including nanosilver (biocidal silver), enhance its efficacy as an antibacterial agent as well as an ecotoxin.

Biological macro-system decline: "Virtually every ecological or biological macro-system on the planet is in decline right now (other than humans and human feed animals, which are both expanding)." (Hartmann 2009, 60).

Biosphere: One planet, one biosphere, one giant ecosystem, many interrelated, integrated ecosystems and food webs. Implicit in the existence of the biosphere are the hydrosphere, cryosphere, atmosphere, and geosphere, of which it is constituted and in which it lives. Thank you, photosynthesis.

BPA: "Canada declares BPA, a chemical in plastics, to be toxic." (Austen 2010).

BPA in the ocean: "Of the about 1 million tons of BPA produced annually, 60 percent of it ends up in the ocean." (Buckley 2010).

Breast milk: "Perchlorate found in breast milk across US." (McKee 2005).

Breast milk versus formula: "It is true that infant formula contains far lower quantities of dioxins, PCBs and organochlorine pesticides than breast milk. However, formula has serious drawbacks that tip the scale against it." (NRDC 2010; http://www.nrdc.org/breastmilk/formula.asp).

Canaries in your local ecosystem: Bats, bees, and butterflies; frogs, falcons, and fish; fetuses, babies, and kids (your children and my children): the harbingers of the Tragedy of the World Commons. Messengers, but what is their message?

Cancer: Cancer to be the world's top killer by 2010 (WHO www.who.int).

Change: "The nation is in the midst of one of its most perilous economic transformations." (Phillips 2006).

Chlorinated dioxins and furans: CDC biomonitoring found chlorinated dioxins and furans in 1,482 of 1,952 people sampled.

Chlorinated water: Brominated species of trihalomethanes occur more frequently in chlorinated public water supplies than in private, domestic well samples.

Chlorine: There is evidence that adding chlorine, a common process in conventional drinking water treatment plants, makes some pharmaceuticals more toxic (Donn 2008).

Clean power?: "China leads the global race to create clean power" but total power generation in China is "on target to pass the US in 2012 – and most of the added capacity will be from coal." (*The New York Times* headline January, 31 2010).

Color of biocatastrophe: The greening, yellowing, and browning of the suburban and urban landscapes of America, followed by those of the aspiring nations of the developing world – the color changes of a globalizing consumer society too stupid not to commit self-initiated genobiocide.

Cookie dough: To avoid E. coli 0157:H7, don't let your kids eat raw cookie dough.

Covenant: The ecology of money is grounded in the covenant of profit that connects the symbiotic ecosystems of global consumer society.

Credit: "As prices rise, the same collateral can support a greater amount of credit." (Soros 2008b).

Culturally induced threats to natural ecosystems: No better way to commit genobiocide than to contaminate the global water cycle with anthropogenic ecotoxins.

Dead zone: "Lost nitrogen [unused nitrogen from industrial agriculture] is the largest contributor to the 'dead zone' in the Gulf of Mexico – an area the size of Connecticut and Delaware combined, in which excess nutrients have caused microbial populations to bloom robbing the water of oxygen needed by fish and shellfish." (Gurian-Sherman 2009b, 1).

Debt: In the United States, personal debt is now 120% of personal income.

Degenerative farming systems: "Chemically based, degenerative farming systems lead to declines in resource abundance and environmental quality, leaving natural systems in worse shape than they were originally by depleting soils and damaging the environment." (www.rodaleinstitute.org).

Demise: "The demise of the American military petroleum industrial complex." (Phillips 2006).

Derivatives washout: "While the combined derivatives (debt) market grew to a nominal value of $270 trillion in 2004, no one knew what they would be worth in a panic." (Phillips 2006, 331).

Dirtiest: "Fossil fuels derived from oil shales and tar sands are the dirtiest of all carbon based fuels." (Gore 2007).

Drugs in drinking water: "Roughly 100 pharmaceuticals have now been identified in rivers, lakes, and coastal waters throughout Europe and the United States." (Hemminger 2005).

Ecology: Ecology, including human ecology, is the transformation, exchange, and discarding of energy by living organisms. Pollution is derived from the dissipation of energy in human ecosystems.

Ecology of money: The cult of profit replaces ethnic nationalism as a motivating component of the modern global economy as a regimented fascist enterprise.

Economic development: The increasingly intensive methods of exploiting natural ecosystems in a world of increasingly complex and centralized human ecosystems are the prime mover of economic development.

End of empire: "The US needs to use Russian, Chinese and European rockets to launch its space satellites." (Phillips 2006).

Epigenomics: "'Epigenomics is where genomics was 30 years ago, when everyone was working on part of the puzzle.' – Peter Jones." (Qiu 2006, 145).

Evolution: In America, the "petroleum industrial complex" (Phillips 2006) has now evolved into the loan shark-consumer complex.

Fallout: Atmospheric deposition of radioactive and chemical fallout is correlated with rainfall events.

Felonious activity: Operating a global economy under the influence of Reaganomics.

Fisheries biomass: "Industrialized fisheries typically reduced community biomass by 80% within 15 years of exploration." (Myers 2003).

Flags and anthems: The anthem of God's chosen people: Keep on spraying, keep on praying, keep on paving, keep on waving.

Food waste: "The U.S. produces about 591 billion pounds of food each year, and up to half of it goes to waste, costing farmers, consumers and businesses hundreds of billions of dollars." (*The Wall Street Journal* 2010, C12).

Frankenstein laboratories: "In the Frankenstein laboratories of Wall Street, banks created new risk products... without mechanisms to manage the monster they had created." (Stiglitz 2010).

Front lines of biocatastrophe: Public health service workers will occupy the trenches on the front lines of biocatastrophe, having the highest risk of exposure to ABRB and other emerging pathogens in the context of rapidly diminishing public resources and rapidly increasing populations of humans in need of health care.

Fuel subsidy: Yet another version of a credit default swap.

Fuel-subsidized agriculture: Re: non-sustainable industrial agriculture; "...all benefits have their costs." (Odum 1983).

Genetics: "The primary goal of the $3 billion Human Genome Project — to ferret out the genetic roots of common diseases like cancer and Alzheimer's and then generate treatments — remains largely elusive." (Wade 2010).

Global depression: "America is in the happy position of being able to write IOUs for purchases of goods and services that destroy American jobs... Since this irrational arrangement is the only means human wisdom has so far devised to avoid a global depression, it is understandable that there should be a conspiracy of policymakers to keep it going. But it cannot go on and on." (Skidelsky 2008, 64).

Global this and that:

   The glass half full: Global consumer society-global commodities market-global transportation systems-global communications systems-global banking-global recreational opportunities-global medicine-global humanitarian relief-global recreational opportunities.

   The glass half empty: Global warfare-global resource depletion-global soil degradation and salinization-global deforestation-global warming-global sea level rise-global climate change-global ecosystem destruction-global food crisis-global chemical fallout-global contaminant pulses-global transportation of emerging and reemerging pathogens-global pandemics-global denial of Biocatastrophe.

Global trading networks: Biocatastrophe will be characterized by sporadic and intermittent, then increasingly systematic, disruptions of global trading networks.

Goldman Sachs revelations: Lifting the shadows of non-sustainable finance capitalism.

Gospel: America as a high-technology-gospel-spreading super power. The gospel has been spread; what now?

Government debt: "The White House estimates government debt accounted for 90 percent of the economy's total output in 2009, up from 70 percent a year earlier." (Bajaj 2009).

Green revolution: The suppression of sustainable indigenous agricultural communities by industrial agriculture in the late 20th century is the progenitor of the rebirth of sustainable ecoagriculture in the 21st century.

Gun industry: "A 1997 CDC study that compared gun-related death rates in 26 industrialized countries among children less than 15 years old found that the gun-related homicide rate among U.S. children was nearly 16 times higher than the rate among children in the other 25 industrialized countries combined; the gun-related suicide rate was nearly 11 times higher; and the unintentional gun-related death rate was nine times higher." (http://www.consumerfed.org/pdfs/youthfa.pdf).

Guns: "Guns are virtually the last consumer products to remain federally unregulated for health and safety." (http://www.consumerfed.org/elements/www.consumerfed.org/file/health/industry.pdf).

Health advisories: "States that have issued health advisories limiting consumption of fish has risen steadily from 27 states in 1993 to 41 states in 1999. A total of 2,073 advisories were issued." (www.epa.gov/waterscience/fish).

Hedge funds make hay: In *The New York Times* OpEd (April 4, 2010).

Herbicidal ecocide: One stateroom among many in the sinking of the Titanic.

High fructose corn syrup: "Consumption of HFCS has increased tenfold since 1974. The obesity epidemic in America's children precisely tracks this trend." (Landrigan 2010b).

High oil prices: "An indirect consequence of high oil prices [is] the increased demand for biofuels, which is resulting in farmland being turned from food production to fuel production, thus making food more costly." (Cribb 2010, 122).

Horses: Numerous horses have died after chewing on wooden fences infused with the same arsenic based pesticide found in the decks and play sets of 70 million homes nationwide." (http://www.ewg.org/node/26238).

Household debt: "Household debt was only 60 percent of income when Reagan took office, about the same as it was during the Kennedy administration. By 2007 it was up to 119 percent." (Krugman 2009a).

Humans as prey: Humans are the prey of the microbial subsystems that ultimately control human population density.

Income shrinkage: Disguised income shrinkage is the result of white collar job losses due to the efficiency of flat-world information technology and globalization.

Industrial Green Revolution: "Based upon the use of heavy chemical fertilizers in irrigation, the industrial green revolution worked only as long as fuel was cheap and water was abundant." (www.rodaleinstitute.org).

Invasive alien species: A concise description of a selection of one hundred of the world's worst invasive species can be located at http://www.issg.org/database/species/search.asp?st=100ss&fr=1&str=&lang=EN, a component of the global invasive species database.

Juggling paper: The displacement of manufacturing by financial services represents the last frontier for the ecology of money in corporate America.

Lesson of PCB: That toxic assets are more than just sub prime real estate loans.

Life stage vulnerability: "Embryo-larval and early juvenile life stages generally are more sensitive to chemicals than are adults of the same species." (George-Ares 2000,1007).

Literary-Liberty interface: "All of the pathologies that have been explored earlier in this book [*The Assault on Reason*] , deception, secrecy, the politics of fear, the appeal of a 'crusade,' and the substitution of raw power for knowledge and logic are on vivid display in our energy and environmental policies." (Gore 2007, 191).

Macroecology: The study of the biosphere as a bowl of microbial soup. You are what you eat, bon appétit.

MAD: The Myth of Mutual Balance and the Reality of Mutually Assured Destruction (MAD) - Hello bacteria, good-by humankind (nobody gets out alive).

Maximum carrying capacity: The maximum carrying capacity of human ecosystems is limited by the amount of available energy and the ability of a globalized consumer society to fund the cost of this energy.

Mercury: "Mercury concentrations are high enough in piscivorous wildlife to cause adverse behavioral, physiological, and reproductive effects." (Driscoll, et al. *Bioscience* January 2007).

Methylmercury: "Metabolism of mercury by microorganisms in sediments creates methylmercury, an organomercurial compound, which can bioaccumulate in terrestrial and especially aquatic food chains." (CDC 2005, 45).

Methylmercury levels in mink and otter: In the Northeast US, the mean level (ug/g) for mercury in the fur of 126 mink tested was 20.7, with a range of 1.78 to 68.5; the mean level in river otters' fur was 18.0, with a range of 1.14 to 73.7 (Yates 2006).

Microbial subsystem: ". . . the key control mechanism in ecosystems is the microbial subsystem that regulates the storage and release of nutrients." (Odum 1983).

Micronutrient malnutrition: Some 923 million people are seriously undernourished with more than 2 billion people suffering from micronutrient malnutrition or hidden hunger caused by inadequate and non-diversified diets (FAO SOFI Report 2002; www.rodaleinstitute.org).

Mimics: "PCBs mimic estrogen and interfere with thyroid hormone." (http://www.nrdc.org/breastmilk/glossary.asp.)

Missing: What the research culture of modern academia doesn't analyze: the dynamic interrelationships of the social, economic, political, and ecological components of biocatastrophe.

Mitochondria: The location of selected lipid degradation processes, where absorbed, and adsorbed teratogenic, mutagenic, carcinogenic, and endocrine disrupting ecotoxins have the potential to cause long-lasting health effects.

Modern evolutionary processes: "Modern evolutionary processes, as is well known, occur at a faster rate than those of the past." (Tainter 1990, 214).

Mortality: The die-off of large populations is the natural and expected consequence of the demise of global military/industrial/consumer society.

Myopic worldview: The belief that about 20% of flat-world technology beneficiaries will live in a world of "happy chaos" – the most optimistic scenario for an imploding global economy with collapsing ecosystems viability that will characterize the age of biocatastrophe (Starobin 2009).

Nanodada: Nanodata dadadata, i.e. the wonderful world of nanotechnology and its hemisphere-wide proliferation of nanotoxins. Only biotechnology has a greater potential to facilitate genobiocide.

Narcissistic global consumer culture: If you were unnerved by the financial panic of 2008-?, wait until incipient biocatastrophe becomes full-fledged biocatastrophe.

Never ratified (again and again): Basel Convention on the Control of Transboundary Movements of Hazardous Wastes and Their Disposal, May 5, 1992.

New book titles:
- *Onslaught of the Disenfranchised: (Where did all the Ammunition Go?)*
- *The Browning of America: FOX News and the Hegemony of Tea Party Fascism*
- *Hot, Crowded, Polluted, Infected, Genetically Modified, Hormonally Disrupted, Incontinent, Feminized/Masculinized, Unemployed, Uninsured, Uninformed, Manipulated, Marginalized, Indebted, Disenfranchised, Powerless – and Angry*

Nitrogen fertilizer: "Ninety-seven percent of the world's nitrogen fertilizers are made using the hydrogen from natural gas. Availability of industrial N fertilizer is thus limited by the extent of natural gas reserves and their price; moreover, the International Energy Agency has warned that global oil and gas will peak within the decade 2010-20. From this point on, industrially produced nitrogenous fertilizers will become increasingly scarce and expensive, posing a threat to crop yields worldwide unless alternatives are quickly developed." (Cribb 2010, 75).

Nuclear accident: In case of a nuclear accident, to measure the radioactive surface contamination in your organic garden from the atmospheric deposition of biologically significant cesium 137 in becquerels per square meter (Bqm^2), divide picocuries/m^2 by 27. Don't forget to take your boots off when you go in your house.

Obsessive compulsive disorder (OCD): A condition wherein the person is unable to control urges to perform certain activities, often repetitiously, e.g. investment bankers who made wads of money by being addicted to the habitual packaging and resale of collateralized debt obligations (CDOs). Packaging groups of worthless subprime mortgages, OCD bankers resold these derivatives as highly rated AAA bonds to gullible investors, who in turn purchased insurance from other commercial banks (e.g. AIG) in the form of lucrative, interest-paying credit default swap (CDS) certificates of investment, i.e. toxic assets of the nth order of magnitude.

Ogallala aquifer: "The huge Ogallala aquifer, which underlies eight states in the American Midwest and is extensively used to grow food, is being depleted at ten times the rate of natural recharge, and some experts fear it could dry up completely within twenty-five years." (Cribb 2010, 42).

One biosphere-one chemical monitoring database: The biogeochemical cycles of the biosphere are hemispheric in their patterns of circulation; contaminant signals from anthropogenic chemical fallout are subject to global transport mechanisms. What goes around comes around.

Onset: The worldwide proliferation of ecotoxins is a primary component of the onset of the Age of Biocatastrophe. As a biogeophysical event, its social impact results from the systematic loss of biodiversity and productivity and the consequential collapse of ecosystems viability. The concomitant phenomenon of multiple global financial crises renders the role of this proliferation of environmental contaminants in the etiology of Biocatastrophe as all but invisible.

Organic diets: "Preschool children in the Pacific northwest eating a conventional food diet had 8 times the organophosphorus pesticide exposure compared to children of parents who provided organic diets." (Curl, et al. 2003; Lu, et al. 2005; www.rodaleinstitute.org).

Perfluorocarbons (PFCs): Two PFC derivatives, perfluorooctanesulfonic acid and perfluorooctanoic acid, have been found to be persistent in the environment and are detected in blood samples all over the world.

Personal care products: "There is a growing concern about the potential impacts on aquatic ecosystems of personal-care products and pharmaceuticals such as birth-control residues, pain killers, and antibiotics. Little is known about their long-term impacts on human or ecosystem health, although some may be endocrine disruptors." (UNEP 2007, 135).

Pesticide poisoning: Around the world, 1 to 5 million farm workers are estimated to suffer pesticide poisoning every year, and at least 20,000 die annually from exposure, many of them in developing countries (World Bank July 9, 2007).

Petri dish: Biocatastrophe is human civilization mimicking microorganisms in a Petri dish – death by over utilization of nutrients.

Pharmaceuticals: More than 100 different pharmaceuticals have been detected in lakes, rivers, reservoirs and streams throughout the world, including Asia, Australia, Canada, Europe, Swiss lakes, and the North Sea (Donn 2008).

Pharmaceuticals in livestock: United States livestock farmers dispense 11,000 metric tons of antimicrobial medications every year to promote the growth of animals (Shah 2010).

Pharmaceuticals in wastewater: "It is generally established that about 95% of the pharmaceuticals entering the sewer plant cannot be controlled. The pharmaceuticals enter the wastewater treatment plant… in either the effluent or the sewage sludge biosolids." (Lubick 2008).

Phytoestrogens: "The concentrations of phytoestrogens detected in the blood of infants fed soy formula were 13,000 to 22,000 times greater than the concentrations of natural estrogens." (NRDC 2010; http://www.nrdc.org/breastmilk/formula.asp).

Poisoned wells: "The wells of the financial system have been poisoned." (James Galbraith November 25, 2008 *MPBN*).

Pollen allergen sources: Increasing CO_2 emissions accelerate pollen allergen sources.

Provincial woe: "Woe to those in the provinces, on the margin of things, and unable to think or operate in a global way." (Starobin 2009, 322).

Race: "The race to save the international financial system is still ongoing." (Soros 2008b).

Radioactive boar: "Many of the boar that are killed land on the plates of diners across Germany, but it is forbidden to sell meat containing high levels of radioactive caesium-137 -- any animals showing contamination levels higher than 600 becquerel per kilogram must be disposed of. But in some areas of Germany, particularly in the south, wild boar routinely show much higher levels of contamination. According to the Environment Ministry, the average contamination for boar shot in Bayerischer Wald, a forested region on the Bavarian border with the Czech Republic, was 7,000 becquerel per kilogram. Other regions in southern Germany aren't much better." (Hawley 2010).

Radiation reduction: "A chemical mixture known as Giese salt, when ingested, has been shown to accelerate the excretion of the radioactive substance [caesium-137]. Giese salt, also known as AFCF, is a caesium binder and has been used successfully to reduce radiation in farm animals after Chernobyl." (Hawley 2010).

Radioisotope thermoelectric generator (RTG): Over 1,000 RTGs were used by the USSR for lighthouses and many are unaccounted for, a situation that could potentially be exploited by terrorists to create a dirty bomb.

Rain showers: "Every rain shower is a chemical fallout event." (Capt. T.).

Rainfall event: Rainfall maximizes chemical fallout events by absorbing or adsorbing airborne gaseous or particulate ecotoxins. The discovery of PCBs in Antarctic sea birds in the 1960s by Swedish researchers (Miller 1970) was the first indication of the extent of the global tropospheric transport of persistent organic pollutants (POPs) in the atmospheric water cycle.

Resource shortages: Resource shortages reduce industrial output and the temporarily stable state it generates.

Savings rate: "As of 2005, China had a savings rate in the 43 percent range, while the savings rate in the United States had turned negative – savings were being drawn down." (Phillips 2006, 381).

Sectarian ideal: "Preservation of the integrity of global life support systems." (Odum 1983).

Shopping: I shop, therefore I am (Phillips 2006).

Southern fried: In *American Theocracy*, "Mr. Phillips paints a chilling picture of American religion as a mixture of biblical idealism and southern fried bigotry." (*The Economist* 2006).

Sport utility vehicle: "The amount of grain needed to fill the tank of a sport utility vehicle just once would feed a poor person for a year, the World Bank notes." (Cribb 2010, 125).

Standing still: "The US needs nearly one trillion dollars of foreign money each year just to stand still." (Bloom 2006).

Sustainable cities: "If cities are to be sustainable, extensive supporting ecosystems must continue to provide renewable resources." (McGraw-Hill 2002, 611). The rapidly accelerating world water crisis is now the greatest threat to the sustainability of urban environments (ET).

Tea Party nitwits: The best friend of the Wall Street casino.

Technological optimists: "The hopes of the technological optimist center on the ability of technology to remove or extend the limits to growth of population and capital." From the Club of Rome report (Meadows 1972).

Technometabolism: The seemingly limitless technometabolism of global military/industrial/consumer society will eventually be constrained by the limiting factor of finite, nonrenewable fossil fuel resources.

The hallucinations of the Tea Party fascists: "Terrorism isn't caused by poverty, poverty is caused by terrorism. Terror is a tool used by those seeking power to keep the masses in need of an answer." (Beck 2005).

Tools: Are our tools and technologies the message in the bottle?

Toxic waste: "Toxic waste is the limiting factor for industrial society." (Odum 1983).

Trans Pacific air pollution: "Large scale pollution and dust from Asia … is transported at mid-altitudes across the Pacific Ocean in a matter of days and impacts sites in North America…" (www/esri.noaa.gov/gmd).

Two dollars a day: More than 2.8 billion people, close to half the world's population, live on less than the equivalent of $2/day (waronstarvation.blogspot.com/2008/01/starvation-poverty-statistics.html).

Volatile times: "Markets routed as worry grows on Europe debt… fears that continent's problems may hurt global economy." (*The New York Times* headline February, 5 2010).

Water scarcity: "If present trends continue, 1.8 billion people will be living in countries or regions with absolute water scarcity by 2025, and 2/3 of the world's population could be subject to water stress." (UNEP 2007, 116).

Words and Concepts to Think About

Biocatastrophe Potpourri[1]

American grain-based ethanol production disaster: Increased fuel production for internal combustion engines = decreased food production for hundreds of millions of world citizens.

104

Artificial carbon filtering trees: a new carbon capture and storage technology.

Asset deficit disorder: The envy of other people's belongings.

Autism spectrum disorders: The etiology of autism spectrum disorders is linked to the impact of endocrine disrupting chemicals, which trigger "on and off genes." (http://www.researchautism.org).

Baconian (Francis Bacon) world view: The theory of a mechanical universe based on objective knowledge; a primary theoretical basis for the imposition of human ecosystems on natural ecosystems.

Biometallurgy: The production of rare metals from dirt by designer bacteria.

Cadmium telluride (CdTe): The toxicity of the life cycle of nanoscale CdTe quantum dots used in thin film photovoltaics production.

Carbon crisis: "The energy crisis and the climate crisis are inextricably linked." (Gore 2007, 191).

Cartesian (René Descartes) world view: The mathematical model of a mechanical universe became the architectural framework for the ideology of imposing human ecosystems upon natural ecosystems.

Catastrophic loss of fish species: "If a new analysis of marine ecosystems data is correct, commercial fish and seafood species may all crash by 2048." (Stokstad 2006, 745).

Chlorinated congeners: "Highly chlorinated congeners are more resistant to metabolic breakdown and they tend to be conserved and more reflective of bioaccumulation differences." (Alaska Marine Mammal Tissue Archival Project; http://www.absc.usgs.gov/research/ammtap/ammtap.htm).

Cities as parasites: Very tall slime molds.

Cliff: The alternatives to falling off a biocatastrophe cliff: sliding down the slope of the gradual decline of the viability of global consumer society or, in the best case scenario, achieving a plateau of sustainable economic activity (Starobin 2009).

Closed system: In the closed system of the Earth's biosphere, material entropy must ultimately reach a maximum (Rifken 1980).

Collapse of industrial society: "Industrial societies are subject to the same principles that caused earlier societies to collapse." (Tainter 1990, 216).

Competition-coexistence interactions: The roots of global warfare.

Diesel fuel: "To grow the grain to make one tankful of diesel fuel requires 330 U.S. tons of water." (Cribb 2010, 125).

Disenfranchisement: The disenfranchisement of informed citizens by sectarian ideology and media propaganda.

Dolphins: Dolphin morbillivirus epizootic resurgence.

Ecotoxins as secondary metabolites: Second, third, and ad infinitum generations of toxic assets.

End: The end of the American empire.

End of life hazards: Full life cycle accountability includes end of life hazard ecotoxin abatement; a major challenge of solar PV industries, petrochemical nanomanufacturing, and global consumer society and personal care products.

Endocrine disrupting chemicals (EDCs): Synthetic organic chemicals that disrupt hormone signaling and alter gene expression.

Endocrine disrupting chemicals (EDCs) II: Xenobiotic estrogenicity of individual consumer products and industrial chemicals.

Endocrine disrupting chemicals (EDCs) III: The transfer of synthetic chemicals into maternal cord blood and the subsequent effects on the developing fetus.

Entropy: Dissolution of the productivity of finance capitalism.

Epigenetic codes: Changes in gene expression caused by the cross-placental transfer of endocrine disrupting chemicals (EDCs).

External costs: The hidden costs of profit driven markets

Financial cancer: Global consumer society indebtedness.

Financial innovations of a predatory commercial banking system: Ms. *Collateralized debt obligation* (CDO) marries Mr. *Credit default swap* (CDS).

Food for thought: Hotel California's smorgasbord of ecotoxins, a metaphor for the finite biogeochemical world in which we live. We are what we eat; bon appétit yet again.

Food stress: High prices, low income.

Fuel-powered human ecosystems: The Tragedy of the World Commons - one biosphere, one giant ecosystem: the crisis of accelerating CO_2 emissions obscures the ongoing contamination of the global atmospheric water cycle on a grand scale.

Gas hog culture: Gospel of high technology.

Genetic engineering: The repackaging of genes as a goal of for profit biopharmaceutical corporations.

Genetically engineered (GE) plants: The potential harm of the engineered traits of GE plants as pharma crops.

Genetics: The overwhelming evidence of the link between genetics and autism spectrum disorders.

Glass half full[-3]: The isolation of high income individuals from ecotoxins in foods and consumer products: money is green.

Global bleaching of coral reefs: An early indicator of ecological distress.

Global financial markets: Anthropogenic economic ecosystems are characterized by increasing complexity and centralization, and, thus, by vulnerability to disruption and collapse.

Global mass flora and fauna extinction event: Where did the whip-o-wills in my yard go?

Globalization: The expansion of the middle class in developing countries.

Globalization of disease: Global trade, global transportation, global illness.

Green slime: The relationship of green slime blooms in marine environments to nitrogen pollution from industrial agriculture runoff.

Growth variance: "Undifferentiated growth vs. multi-level organic growth." (Odum 1983).

Halogenated flame retardants: Among the most pervasive of all Wall Street ecotoxins.

Hedge funds: Cascading cross defaults.

Human ecology: The cause of the mother of all mass extinction events, genobiocide.

Hydrofracking: The high external costs of fossil fuel recovery, in this case natural gas, will be the cause of widespread contamination of underground aquifers.

Illusions: Beware the "illusion of happy chaos" (Starobin 2009).

Implosions: Implosions of American manufacturing as predictor of the implosion of the American economy.

Impostor hormones: Non-hormone molecules which as endocrine disrupting hormone imitators block or alter the function of natural hormones.

Industrial agriculture: A prime mover of the destruction of local forms of knowledge.

Instability: "Structural instability of our financial institutions." (Shiller 2010).

Intersex: The feminization of male fish due to endocrine disrupters in marine and lacustrine food webs.

Irony: Slightly green Tea Party fascists.

Joint compounds: A major source of PCB emissions, as in airport runways, highway bridges, and other building material uses.

Land degradation: Land degradation as a feedback mechanism for global warming and cataclysmic climate change.

Lipophilic: Fat loving ecotoxins, due to their solubility in living tissues, bioaccumulate in pathways to human consumption.

Marginalization: The marginalization of traditional religious groups and the radicalization of US Protestantism (Phillips 2006).

Microbial epidemics: The proliferation of ABRBs (antibiotic resistant bacteria) and reemerging or newly emerging viral infections will be the most important organisms in a future world of ever changing ecology.

Microcommunity: The vast, invisible, interrelated communities of bacteria, fungi, and other microorganisms that will eventually recycle all biotic material including human corpses.

Minor detail: Increasing death rates from drug resistant infections.

Mixed grill: Today's special, funky assets: Reaganomics credit binge, legacy costs, and credit default swaps for dinner, ecotoxins for dessert.

Monkey pox: One of the many zoonotic viruses impacting human health.

Monsanto: Biopiracy of genetic information.

Myth: American exceptionalism and the myth of endless prosperity.

Neonicotinoids: Nerve gas insecticides that have replaced organophosphates and carbonate pesticides. The name of the pesticides implicated in beehive collapse syndrome.

Neurotoxic dust: Ubiquitous chemical fallout from the commonplace, convivial, electronic equipment of the age of information technology.

Neurotoxins: The labyrinth of relationships between neurochemicals, neurotransmitters, mirror neurons, and endocrine disrupting chemicals (EDCs). Hello, methylmercury, where are you today?

Non source-point pollution: Chemical fallout ecotoxins, originally manufactured at specific industrial source points, incorporated in and widely disbursed in innumerable industrial and consumer products, and then transported as global contaminant pulses due to their volatility or solubility and tendency to be continually recycled within the biosphere.

Oil based society: "The side effects of an oil based society… are economic volatility, geopolitical conflict, and the climate damaging impact of hydrocarbon pollution." (Roberts 2005).

Ominous biomarker: "Levels and trends of polybrominated diphenylethers and other brominated flame retardants in wildlife" (Law 2003).

Pollution: Waste as dissipated energy (Rifkin 1980).

POPs: The preternatural transfer of POPs from adult female marine mammals to the fetus.

Power-elites: The dis-informed special interest power elites.

Primordial transport vector: The ingestion of ecotoxins by microorganisms, which then synthesize them within or on the surface of amino acids.

Privatization of the Commons: The generous public donation of U. S. Patent and Trademark Office (PTO) generated innovations and inventions to "private institutions of the commercial control over resources and ideas that are part of the shared commons." (http://www.icta.org/patent/index.cfm; 2010).

Protein: Increasing global per capita protein intake and its impact on agricultural productivity.

Reproductive toxicity: Adverse effects of ecotoxins on the neurological, immunological, and physical development of infants and on sexual function and fertility of adults.

Say it again: "The race to save the international financial system is still ongoing." Repeat George's portent ten times, recite the Hail Mary, and retire for a very long evening.

Sea level rise: The rate of sea level rise expected by the end of this century, as calculated by expert climatologists, has almost doubled between 2007 and 2009 (Wilkinson 2009).

Silicon chip production: The nano and pico technologies of silicon chip production generate a number of potent greenhouse gasses as well as a wide variety of potent powdered ecotoxins, some of which have not yet been identified.

Soil bacteria: The role of soil bacteria in producing soil nitrogen.

Survival: The literacy-liberty interface.

Switchgrass: The most useful of all biofuels (except algae??)

Synergism: The synergism of historical events is the limiting factor for the continued growth of human ecosystems.

Synthia: The emerging synthetic biology industry.

Therapeutic society: One in which professional elites medicalize acts of will and minimize personal responsibility (Lasch 1977).

Too late: The greening of the economic elite.

Too stupid to punt: A consumer society that can't adapt to the end of the age of oil by adapting renewable energy sources.

Toxic assets: Subprime food, water, agricultural produce, oceanic fisheries, forests, soil; subprime real estate is just the tip of the toxic assets iceberg.

Toxic footprints[$]: The worldwide ecotoxic footprint of the techno-elite.

Tragedy of the World Commons Topic 3: The undermining of indigenous sustainable agricultural communities by industrial agriculture.

Tragedy of the World Commons Topic 22: Unintentional chemical byproducts.

Transnational corporations: The collective expropriation of the assets of the World Commons.

Twenty first century epiphany: The broad recognition of the non-sustainability of an industrial civilization based on nonrenewable supplies of fossil fuels.

Ultimate geography lesson: The biogeography of anthropogenic ecotoxins.

Ultra-trace unwanted byproducts: The future legacy of The Age of Nanotechnology.

Unsolved mysteries: How ubiquitous environmental chemicals interfere with the functioning of membrane and nuclear hormone receptors to control cell processes.

Unsustainability[2]: When the participants in global consumer society run out of credit.

Unsustainability[3]: When the institutions of finance capitalism can't service their own debt.

Unsustainable input of resources: The key element in the collapse of the infrastructure of an increasingly complex network of the industrial activities of a globalized consumer society.

US debt danger zone: National debt + household debt + commercial debt + financial sector indebtedness + foreign current account deficits (Phillips 2006, 338).

Wal-Mart Woe-Man (WMWM): The universal, unisexual customers of the global market economy, characterized by the habitual use of credit cards to purchase consumer products, a high indebtedness, an obliviousness to the functioning of corporate plutocracy, and a minimal awareness of the nature of his/her ecotoxic legacy to their grandchildren.

Water: The limiting factor for human existence; since the beginning of the age of chemical fallout (1945), the atmospheric water cycle has been contaminated by over 80,000 anthropogenic chemicals. What now?

World Bank: The sponsor of organized corporate crime in the form of the systematic displacement of sustainable local and regional agricultural and economic activities and enterprises by predatory multinational entities.

Biocatastrophe Potpourri[2]

A shadow banker's best friend - a Tea Party nitwit

Abrupt climate change event

Acute aquatic toxicity

Agricultural biotechnology industry

Agroecological farming

American assumption of entitlement

American automobile culture

American petropolitics

Anthropogenic ecotoxins in marine mammals

Antibiotic resistant gonorrhea

Antibiotics in confined animal feeding operations (CAFO) and their relationship with growing antibiotic resistant infections

Antimicrobial resistance

Anxiety of the electronic elite

Asset bubbles

Azodicarbonamide in plastic seals

Bioaccumulative contaminants

Bioavailable estrogen

Biocatastrophe as the penultimate black swan

Biofuels disaster

Biogas generators

Biopiracy

Biosecurity in poultry production

Biospecimens

Biotech burger

Biotoxins

Black swans

Bottom trawling

Browning of the greens

Bundled and securitized subprime assets scam

California as the after America landscape

Cam-based automation

Capture fishing versus aquaculture

Cascading impact of Gulf oil disaster

CCD (bee colony collapse disorder)

Cheat foods

Chemical burns from antifungal agents in Chinese sofas

Chemical fallout in wood ash

Chemical mutagenesis

Chemical/genetic interface

Christian otherworldliness

Client states of the global plutocracy

Climate change refugee

111

Climate disruption

Closed system aquaculture

Coal tar sealants on driveways and parking lots as important ecotoxin nonpoint sources

Collapse of the casino culture

Collateralized debt obligations

Community associated MRSA

Coral reef extinction

Cord blood biomonitoring

Corporate mismanagement of natural resources

Corporate plutocratic kleptocracy

Credit-industrial complex

Crop diversity

Culture of illusion

Cumulative destabilization of the Middle East

Current account deficits

Cyberterrorism

Dawn of ocean energy

Dawn of the super interglacial drought

Debt as the Achilles heel of American finance

Debt culture of petropolitics

Deepwater Horizon oil spill as a regional biocatastrophe

Derivative free rein

Developmental neurotoxins

Diminishing global natural resources

Diphenol ether congeners

Dire state of many of the world's fish stocks

Dirty capitalism vs. clean capitalism

Discarded resources as pollution

Drug resistant super bug

Duration of the Gulf oil spill disaster

E. *coli* 0157:H7

Eating disorders as an external cost of processed foods

Ecological credit crunch

Ecological logic of capitalism

Ecotoxins in maternal cord blood

Electro dollars and petro dollars

Elite medical network

Emaciated government

Emerging contaminants

Emerging discontinuities of the corporate plutocracy

Emerging global Jihad against western consumer culture

End of the Age of Antibiotics

Endocrine disrupters in sewage sludge

Endocrine disrupters that mimic hormones

Energy internet

Energy Pearl Harbor

Energy privatization

Engineered nanoparticles

Environmental costs of shale rock blasting

Environmental endocrine hypothesis

Environmental tipping points

Epigenetic chemical modifications

Equitable distribution of potable water and whole foods

Estrogenic activity of industrial chemicals

Estrogenisation of the environment

External debt vs. currency reserves

Eye, ear, and throat MRSA

Falling asset prices

Fetal exposure to EDCs

Flashmobsters of the shadow banking network

Fiber optic infrastructure

Financial engineers

Food resource scarcity

Food sovereignty movement

Forgotten entropy

Fungal infections in hatchery fish eggs

Gas flaring as a waste of a nonrenewable energy source

Genetically engineered fish

Genetically engineered Roundup ready alfalfa

Genetically engineered super germs

Geoengineering

Global biocapacity

Global climate catastrophe

Global commons

Global currency crisis

Global economy as a giant feedback loop

Global warming footprint

Globesity

GMO terrorists

Gram-negative bacteria

Great Pacific garbage patch

Growth and collapse behavior mode

Guerilla farmers and gardeners

Heat stress

High frequency trading using supercomputers

High speed electronic trading

High-IQ risk-takers

Homo economicus

Human body water cycle toxicity

Human fetal liver tissue as a bioindicator

Hydroxylated metabolites of PCBs and other chlorinated phenolic compounds

Hyperpartisanship

In utero exposure

Industrial production of finfish

Industrial society as a threat to humanity

Industry's secret chemicals

Inevitability of a debt induced crisis

Invasive alien species

Inverted totalitarianism

Investment globalization

Knowledge as a commodity

Knowledge-based economies

Land degradation

Large-scale resource assessments

Layers of hidden toxic assets

Leveraged investments

Lice alert in Norwegian salmon

Long term capital management (LTCM) fiasco

Long wave events

Loss of ecosystems services

Low concentration exposure

Malfunctioning genetic switches

Malnutrition

Margin calls

Marginalization of the educated elite

Marine debris

Mayflies as bioindicators of fresh water quality

Meltdown in federal government finances

Metabolic pathways

Metallo-activates

Methane-caused warming

Methylmercury-induced toxicity

Microbial fuel cells

Micronutrient deficiencies

Microplastic nanopollutants

Mortgage-backed securities

Multiple track economy of flat-world economics

Multipolar financial world

Nano-agrochemicals

Nano-particle-size plastic antibodies

Nanobiotechnology monitoring systems

Nanomaterials

Narcissistic Tea Party nitwits

National disenlightenment

Neurodevelopment disorders

Nontrivial possibilities

Nutrient pollution

Nutrient risk assessment

Obsolete energy technology

Off balance sheet debt

One way trade (oil)

Outgreening

Parasitic neurological diseases in lobsters

Peak water, soil, land, oil, fish, and agricultural productivity

Perchlorate-induced thyroid dysfunction

Petro Imperialism

Pharmaceutical residues

Pharming

Prevalence of bacterial and fin diseases in farm-raised fish

Privatized profit-socialized risks

Problem of growth in a finite system

Public health advocacy

Recursion feedback loop

Red palm beetle

Regenerative organic farming

Regional food economy

Replacement of manufacturing by finance

Rich poor gap

Riptide of plutocratic influence

Rising burden of neuropsychiatric disorders

RNA interference

Roundup laden foods

Roundup Ready

Roundup resistant pigweed

Roundup resistant super weed

Runaway greenhouse effect

Sea of estrogens

Seafood advisory levels

Secondary organic aerosol particles

Secular dictatorship

Security apparatus of the global warfare state

Self adjusting element of global financial collapse

Self reinforcing events

Semicarbazide in baby foods

Sewage sludge biosolids

Shark fin soup

Siberian shelf methane chimney

Simultaneous exposure to multiple common EDCs

Smart Grid

Social dislocation

Social side effects of new technologies

Social trust

Socially created vulnerabilities

Soil nutrient degradation

Sources of urogenital effects

Species-depleting shell thinning

Spermicidal corn

Stratospheric sulfates for global cooling

Structural instability of our financial institutions

Structural unemployment

Subprime nation

Subprime oceanic fisheries

Super bug crises

Super organism

Sustainability of long term borrowing

Sustainable seafood

Synthetic financial instruments

Synthetic nanoparticles

Technoculture

Technology of mass persuasion

The cyber magic of electronic transfer

The end of the industrial green revolution

The hope for "reality-based economics"

The triumph of utopian economics

Transgenerational effects

Transplacental carcinogens

Unequal patterns of consumption

Unilateralism

US oil vulnerability

Utopian economics

Volcanic dust – jet engine interface

Water mining

Water refugees

White nose syndrome in bats

Wireless nano surveillance systems

Worldwide subprime food crisis

Worldwide topsoil losses

Questions to be Answered

Accidental extermination: If genocide is the deliberate extermination of a nation or people, what is the appropriate word to describe the accidental extermination of a large percentage of humanity and other living species by often well-intentioned human activities?

Biomonitoring: Given the global crisis of rapidly growing individual, corporate, and governmental indebtedness, will it be cost effective to reduce rather than increase expenditures to facilitate biomonitoring for hazardous chemical fallout in pathways to human consumption and as body burdens in human populations?

115

Biotoxins: What was the toxins emissions profile (TEP) of US Army burn pits in Iraq and Afghanistan?

Casino culture: In a CNN commentary on March 27, 2009, the pundit David Gergin, used the term "casino culture" to describe the recent evolution of America's shadow banking network. Is there any chance the casino culture nurtured by our financial engineers, the legacy of America's hang loose unregulated free enterprise market economy, will be the key component in the eventual collapse of global consumer society?

Collapse of complex societies: Will the end of Chimerica coincide with the termination of the use of the dollar as the world's reserve currency?

Collateralized debt: Who made the profits from the sale of worthless collateralized debt repackaged as AAA bonds?

Collateralized debt obligations: Why would investors purchase stocks and bonds secured (collateralized) by worthless subprime real estate loans?

Credit default swap I: If the shadow banking network issued 74 million credit default swap insurance policies in 134 countries, what is the total value of these policies, what percentage will default in the next decade, and what is the total liability of the shadow banking network?

Credit default swap II: What financial institutions now hold the "mother of all toxic assets" i.e. credit default swaps? To avoid the complete collapse of the American banking system, won't American taxpayers own an increasing percentage of these and other categories of toxic assets as federal, corporate, and private debt increases in the future?

Current account imbalance: If American upper and middle class conspicuous consumers, along with lower income Wal-Mart Woe-Man, owe $2.5 trillion in credit card and consumer debt, $16 trillion in national debt, $75 trillion in unsecured corporate and real estate debt, and trillions more in legacy obligations (pensions, social security, Medicare, Medicaid), who will repay this current account imbalance and how?

Dollar: How long before the dollar will cease to be the world's principal reserve currency?

Earthworms: What is the difference between the ecology of earthworms and the ecology of money? (Hint: nutrient element recycling vs. the depletion of the natural resources of the World Commons.)

Entrepreneurial techno-elite: Will the inventive innovative entrepreneurial techno-elite discover sustainable solutions to the challenge of human survival in the 22nd century?

Euro debt: Which banks hold the 2.6 trillion dollars of debt issued by the end of 2009 by public- and private-sector institutions in Spain, Greece, and Portugal?

First responders: In the post-apocalypse era of the age of biocatastrophe, who will be the first responders when whole foods and potable water supplies run out, and in what communities will they reside?

Food miles: How many miles did your (and my) cantaloupe travel before arriving at your (and my) kitchen table, how many calories were expended to transport it, and what are the external costs of the consumption of this cantaloupe?

Food security: How secure are the sources of food in your community in a world of global industrial agricultural systems? If you suffer from "food insecurity," is it because you no longer have the financial resources to feed yourself and your family in a world of ever widening income disparities?

Genetic change: What is the relationship between intake of EDCs by a mother prior to childbirth and the etiology of genetic change leading to the rapid increase in autism spectrum disorders?

Genetic manipulation: What are the unexpected health physics impacts of modern biogenetic manipulation of bacteria for gene coding for the purpose of developing human and animal insulin and growth hormones?

Global market economy: What is the fate of the global market economy in a world of ever-shrinking natural resources, financial assets, and ever-expanding indebtedness?

GMICS (global military/industrial/consumer society): Who will identify and biomonitor the ecotoxins produced by the global military/industrial/consumer society and its global network of the for-profit production of ecotoxins by governments, corporations, and consumers?

Green Revolutions: What is the difference between the first Green Revolution and the second Green Revolution?

Gulf oil spill I: What are the long term environmental impacts of the use of oil dispersants in the Gulf oil spill disaster?

Gulf oil spill II: When will the secrecy surrounding the proprietary constituents of these dispersants be lifted to allow evaluation of their ecotoxicity?

Gulf oil spill III: Will the oil enter the loop current and contaminate the coral reefs of the Florida Keys, and then enter the Gulf Stream, impacting East Coast beaches from Florida to southern New England?

Health care funding: In a world where the global financial crisis results in thousands of billions of dollars being spent on collapsing financial institutions, can any money from parasitic health care insurance company profits and salaries be diverted to reducing income-related disparities in the health care delivery system?

Human health studies: "Could we be trying to correlate exposure and effect at the wrong time? If it is prenatal, or early life stage, exposure that is critical to disease susceptibility, why are we measuring environmental chemicals in people once they have developed breast cancer? The critical exposure window may have been much earlier." (Birnbaum 2003).

Hyperthyroidism: If the chemicals PBDE and BPA cause hyperthyroidism in cats, is there any chance that these chemicals, as potent endocrine disruptors, will cause hyperthyroidism or other health effects in children?

Infrastructure collapse: What communities will survive the Age of Biocatastrophe and flourish again in the post-Apocalypse?

Inorganic bromine: What is the climatological significance of stratospheric inorganic bromine and what are its point sources (www.esri/noaa.gov/gmd/about/ozone.html)?

Jesus Christ Superstar: Where is Jesus Christ Superstar now that we need him to bail us out of the predicament of a contaminated global atmospheric water cycle?

Katrina: What is the inventory of ecotoxins discharged into the Gulf of Mexico from the lower Mississippi River urban areas and chemical plants as a result of the Hurricane Katrina washover event?

Life support resources: Can the biomachinery of synthetic biology save the planet?

List of Lists: What is the difference between the EPA's categories of hazardous chemicals and extra hazardous chemicals? (www.epa.gov/emergencies/docs/chem/title3_Oct_2006.pdf).

Massive die-offs: What do massive die-offs of New England forests caused by the Asian long horned beetle tell us about the fragile and vulnerable situation of our forest resources in a world of declining bat, bee, and bird populations?

Medical industrial complex: What is it, and what is the relationship of MRSA (methicillin-resistant *staphylococcus aureus*) and other ABRBs to its labyrinths of pathogen pathways?

Money: Why do Tea Party fascists and conservative ideologues support the accelerating shift of income to the palisaded elite?

Monoculture: Buy now, pay later; high profits precede the declining productivity of aging monocultures; will the productivity of the global money farms collapse after the next world financial crisis?

MRSA: How long before MRSA becomes resistant to one of a handful of antibiotics of last resort, vancomycin, which still has some utility in combating the spread of this highly infectious ABRB?

Nashville flood: What is the inventory of the biologically significant ecotoxins and pathogens in the untold gallons of untreated sewage, gasoline, fuel oil, and other chemicals washed down the Cumberland River during the Nashville flood of 2010?

Nick Tech: Can Nick Tech of the Technoelite help save communities of the educated elite with electronic inventions, bioengineered bandages, and the cybernetic creation of income-producing information technologies in the context of anthropogenic ecosystems collapse?

Oceanic dead zones: What other ecotoxins are in the toxic content profile (TCP) of oceanic dead zones aside from the nitrate-effluents produced by industrial agriculture?

Penultimate Question: What happens when the global market economy can only grow by the further accumulation of debt?

Pollution absorption: What happens when environmental pollution absorption mechanisms become saturated?

Seventy-five trillion dollars: How far will $750 billion in taxpayer money go to ameliorate AEFFESs accumulation of $75 trillion in credit default swap "toxic assets?"

Socialized banking system: What prompted free enterprise, neo-conservative Republicans to socialize the losses of the American commercial and investment banking system in response to the collapse of the subprime mortgage market? Christian Environmental Fascist Free Enterprise Financial Engineers, where are you now and what did you do with your five star credit default swap certificates?

Storm warning A: How many different ecotoxin contaminant signals can be identified after a *storm washout event* in rainfall draining from mall parking lots, urban wastelands, and industrial agricultural ecosystems into the Pacific Ocean from a southern California rainfall event?

Storm warning B: How many different ecotoxin contaminant signals can be identified after a *storm washover event*, e.g. in the seawater incursion during Hurricane Katrina, as saltwater covering homes, automobiles, factories, waste sites, and asphalt streets drained out into the Gulf of Mexico?

Survival: Isn't the viability of human society dependent on a growing awareness of the necessity of sustainable indigenous agricultural activities?

Sustainable economies: What are the essential ingredients of a sustainable economy?

Technometabolism: When will the increasingly intense biotechnometabolism of industrial society utilizing non renewable energy resources and nuclear fission result in the collapse of the infrastructure of the complex ecosystems of GMICS (global military/industrial/consumer society)?

Titanic event: We have already hit the iceberg, but the majority of passengers on the Titanic planet Earth are unaware of the inevitable outcome of the consequences of our anthropogenic creations. Of a current world population of 6.8 billion people, how many will survive the Age of Biocatastrophe?

Toxins: Do toxins cause autism? (Nicholas D. Kristof, February 25, 2010, *The New York Times*, A33).

Worldwide disease surveillance systems: What will be the health physics impact of the proliferation of antibiotic resistant bacteria (ABRB) and other emerging viral infections as typified by MRSA? How many new forms of antibiotic resistant bacteria or newly emerging viral infections, in what locations, and with how many casualties will occur before improvements in worldwide disease surveillance systems are formulated? What recombinant DNA technology biopharmaceuticals can be engineered by entrepreneurial biopharmacists to combat this onslaught of microorganisms?

Last Questions

The evasion of history: Why is the human community, in the form of modern industrial society and its now globalized market economies, in denial of the consequences of its technological and petrochemical inventions?

Demise of modern industrial society: Is there any chance that the implosion of the institutions of finance capitalism represents the first painful stages of the demise of modern industrial society? How did the sustainable market economies of the age of world exploration and settlement mutate into a predatory global consumer culture based on the unsustainable corporate exploitation of a biosphere with limited resources?

Survival: What happens when the national governments of developing, as well as developed, nations can no longer afford to provide basic social services, such as health care, clean water, adequate food supplies, employment security, public transportation, education, social security for the elderly and disabled, and protection from social unrest?

Post-apocalypse: What inscrutable future will evolve after the collapse of the infrastructure of global military/industrial/consumer society? In the post-apocalypse, will Christian Environmental Fascists, having the benefit of Christological innocence, depart in rapture to their heavenly domain, leaving the rest humanity, including Tea

Party fascists, to survive the turmoil and entropy of the demise of non-sustainable western industrial culture? Will pockets of sustainable communities survive in an era of post global warfare, pandemics, and economic collapse?

Part III – Measurements, Conversion Tables, Acronyms, and Abbreviations

Metric Prefixes

Prefix	Symbol	Multiples	Equivalent
exa	E	10^{18}	quintillionfold
peta	P	10^{15}	quadrillionfold
tera	T	10^{12}	trillionfold
giga	G	10^{9}	billionfold
mega	M	10^{6}	millionfold
kilo	k	10^{3}	thousandfold
hecto	h	10^{2}	hundredfold
deka	da	10	tenfold

Prefix	Symbol	Submultiples	Equivalent
deci	d	10^{-1}	tenth part
centi	c	10^{-2}	hundredth part
milli	m	10^{-3}	thousandth part
micro	μ	10^{-6}	millionth part
nano	n	10^{-9}	billionth part
pico	p	10^{-12}	trillionth part
femto	f	10^{-15}	quadrillionth part
atto	a	10^{-18}	quintillionth part

Common abbreviations for mass, volume, and length measurements

Name	Abbreviation	Value
picogram	pg	1/1,000,000,000,000th of a gram
nanogram	ng	1/1,000,000,000th of a gram
microgram	μg	1/1,000,000th of a gram
milligram	mg	1/1,000th of a gram
gram	g	US penny = ~2.5 grams
kilogram	kg	1,000 grams
liter	L	~0.26 US gallons
milliliter	mL	1/1,000th of a liter
centimeter	cm	1/100th of a meter
millimeter	mm	1/1000th of a meter
micrometer	μm	1/1,000,000th of a meter
nanometer	nm	1/1,000,000,000th of a meter

ppm: parts per million; one part per 1,000,000 parts
ppb: parts per billion; one part per 1,000,000,000 parts
ppt: parts per trillion; one part per 1,000,000,000,000 parts

Units for molar concentration of a solute in a solution

Name	Abbreviation	Concentration	Concentration
millimolar	mM	10^{-3} molar	10^{0} mol/m^3
micromolar	μM	10^{-6} molar	10^{-3} mol/m^3
nanomolar	nM	10^{-9} molar	10^{-6} mol/m^3
picomolar	pM	10^{-12} molar	10^{-9} mol/m^3
femtomolar	fM	10^{-15} molar	10^{-12} mol/m^3
attomolar	aM	10^{-18} molar	10^{-15} mol/m^3
zeptomolar	zM	10^{-21} molar	10^{-18} mol/m^3
yoctomolar	yM	10^{-24} molar (1 molecule per 1.6 liters)	10^{-27} mol/m^3

(http://en.wikipedia.org/wiki/Molar_concentration)

Mass Conversion Factors

→Multiply number of →	by →	to obtain number of
to obtain number of ←	by ←	Divide number of←
mg	10^{-3}	g
g	10^{-3}	kg
g	2.205×10^{-3}	lb
kg	2.205	lb
lb	453.592	g
lb	0.45359237	kg
oz	28.35	g
oz	6.25×10^{-2}	lb
t (tonne = metric ton)	1.00×10^{3}	kg

(Shleien 1998, 2-24).

Commonly Used Reporting Units
Environmental chemicals

mg/kg = thousandths of a gram per kilogram

μg/g = millionths of a gram per gram (parts per million)

μg/kg = millionths of a gram per kilogram (parts per billion)

ng/g = billionths of a gram per gram (parts per billion)

ng/kg = billionths of a gram per kilogram (parts per trillion)

p/g = trillionths per gram (parts per trillion)

μ/L = millionths of a gram per liter

ng/g ww = nanograms per gram wet weight

μg/dL = millionths of a gram per deciliter (1/10 of a liter)

μg/mL = millionths of a gram per milliliter

ng/L = billionths of a gram per liter of liquid

ng/mL = billionths of a gram per thousandths of a liter

Environmental radioactivity

Bq/kg = becquerels per kilogram

Bq/g = becquerels per gram

p/kg = trillionths of a curie per kilogram (parts per quadrillion)

p/g = billionths of a curie per gram

ng/m^3 = nanograms per square meter

Useful Definitions

Becquerel: One becquerel, the newer international unit of measurement, equals one disintegration per second of a radioactive material and is abbreviated Bq. 1 Bq = 2.703 x 10^{-11} ci.

Curie: A curie (ci) is defined as the number of emissions from one gram of radium (Ra226): 3.7 x 10^{10} disintegrations per second. 37 billion disintegrations per second.

Limit of detection (LOD): "The LOD is the level at which the measurement has a 95% probability of being greater than zero. (Taylor, 1987). For most chemicals, the LOD is constant for each sample analyzed. However, for dioxins, furans, PCBs, organochlorine pesticides, and some other pesticides, each individual sample has its own LOD. These analyses have an individual LOD for each sample, mostly because the sample volume available for analysis differed for each sample. A higher sample volume results in a lower LOD and a better ability to detect low levels." (CDC 2005, pg. 459).

Mean: The mean of a list of numbers is the sum of the list divided by the number of items in the list, also known as the average.

Picocurie: A commonly found unit of measurement in the older literature and data reports, the picocurie is a source showing 3.7 x 10^{-2} disintegrations per second or 0.037 DPS – just over 2 disintegrations per minute.

Time (T): With respect to radiological decay, ½ T is the amount of time it takes for the radioactive decay of 50% of any amount of a radioactive substance.

EPA Acronyms and Abbreviations

The regulatory lingo utilized by the EPA to set standards for the evaluation of the health physics impact of anthropogenic and naturally-occurring ecotoxins has an important hidden significance. Ecotoxins are often evaluated for their acute toxicity. The EPA uses the term acute exposure guideline level (AEGL) to signify acute toxicity exposure. The EPA also uses screening values (SVs) that include terminology such as "no observed adverse effect level (NOAEL)" and "no observed effect level (NOEL)". Also used are human health screening values (HHSVs), extremely hazardous substance (EHS), and maximum contaminant levels (MCLs). The latter terminology and its abbreviation are often encountered in water quality analysis reports, for example, with the MCL for chloroform bacteria representing the maximum allowable bacterial content for potable water. Anything above the MCL requires restrictions and/or remediation. Of particular interest is the NOAEL, where a particular ecotoxin may be present in measurable quantities, yet below concentrations that can result in observable public health consequences. These regulations can be used to justify the presence of thousands of manmade ecotoxins in consumer products, public water supplies, and food supplies if present in quantities below the MCL. The fly in the ointment now confronting modern global consumer society is the evaluation of the health physics impact of the synergistic interaction of thousands of ecotoxins, none of which are individually present in quantities that even begin to approach acute exposure guideline levels, let alone the more conservative MCLs, NOAELs, and NOELs.

Other Acronyms and Abbreviations

ABRB: Antibiotic resistant bacteria

ABS: Asset-backed securities

ACHM: Alliance for a Clean and Healthy Maine

ADHD: Attention deficit hyperactivity disorder

ADI: Acceptable daily intake

ADR: Accidental durable remnants

AEFFES: American environmental fascist free enterprise system

AEGL: Acute exposure guideline level

AEIC: Anthropogenic ecosystems infrastructure collapse

AGP: Autism genome project

ALS: Amyotrophic lateral sclerosis, aka Lou Gehrig's disease

AMR: Antimicrobial resistance

AOC: Area of concern

APHIS: Animal and Plant Health Inspection Service

API: Active pharmaceutical ingredient

ATN: Autism treatment network

ATP: Advanced technology product

ATP: Agriculture and Trade Policy

ATSDR: Agency for Toxic Substances and Disease Registry

BANANA: "Build absolutely nothing anywhere near anything." (Friedman 2008, 406)

BaP: Benzo[a]pyrene

BCF: Bioconcentration factor

BIPV: Building-integrated photovoltaics

BNR: Biological nutrient removal

BOD: Burden of disease

BPA: Bisphenol A

BRFSS: Behavioral risk factor surveillance system

BW: Body weight

CAB: Current account balance

CAD: Current account deficit

CAD: Computer-aided design

CAFO: Confined animal feeding operations

CAM: Computer-aided manufacturing

CCC: Cataclysmic climate change

CCS: Carbon capture and storage

CDC: Centers for Disease Control, a US government agency

CDO: Collateralized debt obligation

CdTe: Cadmium telluride

CERCLA: Comprehensive Environmental Response, Compensation, and Liability Act of 1980, commonly known as Superfund

CERHR: Center for the Evaluation of Risk to Human Population

CFC-11: Trichlorofluoromethane

CFC-12: Dichlorodifluoromethane

CFC-113: Trichlorotrifluoroethane

CFS: Chronic fatigue syndrome

CFTC: Commodity Futures Trading Commission (US)

CHARGE: Childhood autism risk from genetics and the environment

CIGS: Copper indium gallium selenide

CIS: Copper indium selenide

CITES: Convention on International Trade and Endangered Species

CLO: Collateralized loan obligation

CLRTAP: Convention on Long-Range Transboundary Air Pollution

CMO: Collateralized mortgage obligation

CNC: Computer numerical control

CSA: Community-supported agriculture

CWA: Clean Water Act

DARPA: Defense Advanced Research Projects Agency

DBCP: Dibromochloropropane

DCA: Dichloroethane

DCE: Dichloroethene

DDA: 2,2-bis (*p*-chlorophenyl) acetic acid

DDD: 1,1-dichloro-2,2-bis (4-chlorophenyl) ethane

DDE: 1,1-dichloro-2,2-bis (4-chlorophenyl) ethylene

DDT: 1,1,1-trichloro-2,2-bis (4-chlorophenyl) ethane

DEET: N,N-Diethyl-meta-toluamide

DEHP: Bis(2-ethylhexyl)phthalate (a plasticizer)

DES: Diethylstilbestrol

DIPE: Diisopropyl ether

DNC: Direct numerical control

DTO: Drug trafficking organizations

DWEL: Drinking water equivalent level

EC$_{50}$: Half maximal effective concentration

ECDC: European Center for the Disease Prevention and Control

EDB: Ethylene dibromide

EDC: Endocrine disrupting chemical

EDSP: Endocrine disrupters screening program

EDSTAC: Endocrine disruptor screening and Testing Advisory Committee

EEZ: Exclusive economic zone

EFSA: European Food Safety Authority

EHS: Extremely hazardous substance

EMEP: Cooperative Programme for Monitoring and Evaluation of the Long-range Transmission of Air Pollutants in Europe

ENU: Ethyl-nitrosourea

EPA: Environmental Protection Agency

EPCRA: Environmental Protection Community Right-to-know Act

ESBL: Extended spectrum beta lactamases

ESRL: Earth System Research Laboratory (NOAA)

ETBE: Ethyl *tert*-butyl ether

ETS: Environmental tobacco smoke

EWG: Environmental Working Group

FAO: Food and Agriculture Organization (UN)

FHC: Finance holding company

FIFO: Fish in fish out ratio

FIRE: Finance Insurance Real Estate

FWACS: Flag-waving American consumer society

GDP: Gross domestic product

GECP: Global ecotoxin contaminant pulse

GEMS: Global Environment Monitoring System

GFA: Grid friendly appliances

GINA: Genetic Information Non-discrimination Act

GMD: Global Monitoring Division (NOAA)

GMICS: Global military/industrial/consumer society

GMO: Genetically modified organism

GPGD: Great Pacific garbage dump

GSE: Government sponsored enterprise

GWP: Global warming potential

HAA: Hormonally active agents

HAB: Harmful algae blooms

HCB: Hexachlorobenzene

HCFC: Hydrochlorofluorocarbon

HCH: Hexachlorocyclohexane

HEFS: High energy farming systems

HFC: Hydrofluorocarbon

HFCS: High fructose corn syrup

HGP: Human Genome Project

HHSV: Human health screening value

HSMI: Heart and skeletal muscle inflammation

IARC: International Agency for Research on Cancer

IBLR: Interbank lending ratio

ICCAT: International Commission for the Conservation of Atlantic Tunas

ICCC: International Conference of Climate Change

ICE: Internal combustion engine

ICSC: International Chemical Safety Cards

IHEP: International Human Epigenome Project

IPCC: International Program on Climate Change

IPCS: International Program on Chemical Safety

IRIS: Integrated risk information system (USEPA)

IRRI: International Rice Research Institute

KP: Kleptoplutocracy

KPC: *Klebsiella pneumoniae* carbapenemase

LBO: Leveraged buyout

LC50: Lethal concentration in air or water at which half the test subjects die

LD50: Lethal dose of an ecotoxin that upon administration causes the death of at least 50% of a group of organisms

LISA: Low-input smallholder agriculture

LOAEL: Lowest observed adverse effect level

LOD: Limit of detection

LOEL: Lowest observed effect level

LORCA: Loss of reactor coolant accident

LRTAP: Long-range transboundary air pollution

MCL: Maximum contaminant level (EPA)

MDG: Millennium development goals

MSA: Mean (original) species abundance

MTBE: Methyl tertiary butyl ether

NAWQA: National Water Quality Assessment Program (USGS)

NC: Numerical control

NDM-1: New Delhi metallo-beta-lactamase

NEPA: National Environmental Policy Act of 1969

NGO: Non-governmental organization

NOAEL: No observable adverse effect level

NOEL: No observed effect level

NORM: Naturally-occurring radioactive materials

NPDES: National Pollution Discharge Elimination System (EPA)

NPDWR: National primary drinking water regulations (EPA)

NPL: National Priorities List (EPA Superfund hazardous waste sites)

NSAID: Nonsteroidal anti-inflammatory drug

NSP: Neurotoxic shellfish poisoning

NUE: Nitrogen use efficiency

OAR: Organization for autism research

OC: Organochlorine pesticide

OCD: Obsessive compulsive disorder

ODD: Obedience deficiency disorder

ODS: Ozone depleting substance

OLED: Organic light emitting diode

OPEC: Organization of Petroleum Exporting Countries

PAH: Polycyclic aromatic hydrocarbon

PAMTA: Preservation of Antibiotics for Medical Treatment Act

PBDDs: Polybrominated dibenzo-*p*-dioxins

PBDEs: Polybrominated diphenylethers

PBDFs: Polybrominated dibenzofurans

PBPs: Pentabromo-phenols

PBTs: Persistent bioaccumulative toxics

PCBs: Polychlorinated biphenyls

PCDDs: Polychlorinated dibenzo-*p*-dioxins

PCDFs: Polychlorinated dibenzofurans

PCE: Perchloroethene

PCNs: Polychlorinated naphthalenes

PCP: Pentachlorophenol

PCTs: Polychlorinated terphenyls

PEF: Private equity fund

PET: Polyethylene terephthalate

PFDA: Perfluorodecanoic acid

PFOA: Perfluorooctanoic acid

PFOS: Perfluorooctane Sulfonate

PIP: Pesticide information profile

PNAS: Proceedings of the National Academy of Sciences

POPs: Persistent organic pollutants

PPCP: Pharmaceuticals and personal care products

PRV: Piscine reovirus

PSP: Paralytic shellfish poisoning

PTO: U. S. Patent and Trademark Office

PUF: Polyurethane foam

PV: Photovoltaic

rbGH: Recombinant bovine growth hormone

rbPRL: Recombinant bovine prolactin

REACH: Registration, Evaluation, Authorisation and Restriction of Chemical substances (European Union)

RGM: Reactive gaseous mercury

ROHS: Restrictions on the use of certain hazardous substances (European Union)

ROS: Reactive oxygen species

RSL: Relative sea level

SARA: Superfund Amendments and Reauthorization Act

SBN: Shadow banking network

SCCPs: Short-chain chlorinated paraffins

SDWA: Safe Drinking Water Act

SEC: Securities and Exchange Commission (US)

SES: Socio-economic status

SHF: Sulfur hexafluoride (SF6)

SONS: Spill of national significance

SOP: Sustainable ocean project

SST: Supersonic transport plane

SST: Sea-surface temperature

STAMP: Seabird Tissue Archival and Monitoring Project

STM: Scanning tunneling microscope

SV: Screening value

SVOC: Semi Volatile Organic Compound

T: Time

TAME: *tert*-amyl methyl ether

TARP: Toxic asset recovery plan

TBBPA: Tetrabromobisphenol A

TCA: 1,1,1-trichloroethane

TCBT: Tetrachlorobenzyltoluene

TCDD: 2,3,7,8-tetrachlorodibenzo-*p*-dioxin

TCE: Trichloroethene

TCP: Toxic content profile

TDI: Tolerable daily intake

TEB: techno-elite bioengineer

TEF: Toxic (TCDD) equivalency factor

TEP: Toxic emissions profile

TEQ: Toxic equivalent

TF: Transfer coefficient

THMs: Trihalomethanes

TIP: Toxic input profile

TPF: Tea Party fascist

TPN: Tea Party nitwit

TPT: Tea Party Taliban

TRI: Toxic release inventory

TU$_a$: Acute toxic units

USEPA: United States Environmental Protection Agency

VOC: Volatile organic compound

VRE: Vancomycin-resistant *enterococci*

WEEE: Waste electrical and electronics equipment

WIP: Work in progress

This text is a WIP. Corrections, suggestions, and additions are welcomed. Email them to CaptainTinkham@gmail.com.

III. Biocatastrophe Pie Charts and Tables

Overview

The following pie charts, tables, and figures are graphical sketches of the phenomenon of biocatastrophe and the ongoing financial crisis of US and global consumer society. The first pie chart summarizes our take on the anthropogenic ecotoxins that are the major biological driving forces of the phenomenon of biocatastrophe. The second pie chart is a more specific listing of some of the most important components of the first chart. The text of this publication provides much more detail about the extraordinary number of ecotoxins discharged into the biosphere by human activity.

The tables on debt, income, and gross productivity are intended to help elucidate the relationship between productivity, world and national assets, and growing world debt. It is the existence of huge national and international indebtedness that has a vast impact on the capacity of modern society to confront and mitigate the consequences of the contamination and destruction of our Round-World Commons with anthropogenic ecotoxins and greenhouse gas emissions, and meet the challenges of proliferating bacterial and viral infections.

This section of the biocatastrophe lexicon ends with two emblematic pie charts pertaining to the trade names of plastics and the performance and compounding data of various plasticizers. These charts provide a suitable introduction to the contents of *Volume 3* of the *Phenomenology of Biocatastrophe* publication series, which begins with a review of the chemistry of ecotoxins, especially those produced by the petrochemical industry. This volume also comments on or documents their toxicity, their ubiquitous presence as industrial wastes, in consumer products, as contaminants in the atmospheric water cycle, and as constituents in biotic media, including maternal cord blood.

Questions

Many questions arise about the economic components of biocatastrophe, especially about the extent and impact of our indebtedness as a nation. Forty trillion dollars is the amount of money frequently noted as the unfunded cost of Social Security, Medicare, and Medicaid. With respect to our current national external and internal indebtedness, how much of these legacy costs, given their decades-long timeframe of obligatory collection and expenditure, should we have already appropriated? In the following table we use 8/40 trillion dollars to indicate the amount of this total that should have been saved by 2009. Another key question pertains to the total actual indebtedness of our commercial and shadow banking networks. If total stock, bond, and other paper assets had a world value of 140 trillion dollars in 2005, what proportion of these assets are now overvalued toxic assets as a result of the global financial crisis, and what percentage of lost asset values are being held by the American shadow banking network? The same question can be asked about the 60 to 80 trillion dollars in credit default swap obligations now held by American shadow banking network investors as well as those in other countries. In view of the 8 to 10 trillion dollars already provided by the government in emergency appropriations, corporate loans, TARP monies, and loan guarantees, how much additional money will be needed to avoid national and global financial collapse, especially if corporate and shadow banking network (e.g. AIG) debt is in the 25 to 50 trillion dollar range, as is likely the case? A more recent emerging controversy is the growing anxiety pertaining to the viability of the Euro currency given the accelerating indebtedness of Greece, Portugal, Spain, and Italy, not to mention England and Ireland. To what extent will this crisis affect the "recovery" (an illusion?) of western market economies from the world financial crisis of 2008? How do the declining prospects of economic growth in European countries and the concomitant issue of a rapidly aging population impact the future stability of the Euro currency and the hope for a return to prosperity? While specific estimates of debt as well as national and world growth and productivity differ widely between online information resources, the reality that now confronts us is that:

✓ The western model of a successfully functioning global consumer society based on finance capitalism and characterized by growing personal, national, and world debt is entering a period of instability and probable implosion.

✓ A key element in this implosion is the production and distribution of military, industrial, consumer product, and information technology wastes and ecotoxins.

✓ The declining supplies of fossil fuels for transportation systems and petrochemical production are a definitive limiting factor for future economic growth.

✓ Sustainable local and national economies are undermined by the continuing spread of chemical fallout throughout the biogeochemical pathways of the atmospheric water cycle and the accelerating production of greenhouse gasses from fossil fuels and other sources.

✓ The accelerating heath physics impact of the proliferation of these bioactive environmental chemicals will impact all human communities, irrespective of socio-economic status.

✓ A continuing increase in the production of nonessential and/or environmentally destructive products for an ever-growing world community of conspicuous consumers is no longer a sustainable option for growth-oriented western market economies, or for the rapidly expanding economies of developing nations, (China, India, South Korea, Indonesia, Brazil) who choose to emulate the model of western market finance capitalism.

132

- ✓ The current global financial crisis has an intricate interrelationship with the phenomena of environmental degradation, chemical fallout, and the proliferation of bacterial and viral infections that characterize complex industrial society with rapid global transportation technologies.

- ✓ If the threats to the World Commons posed by the advent of the age of biocatastrophe are to be mitigated, all components of the impact of human activity on the biosphere must be addressed as a unified phenomenon.

- ✓ The survival of complex human societies will only occur with a return to sustainable economies, energy systems, and consumer lifestyles. The model of an ever-growing global consumer society devouring ever larger quantities of non-essential goods has been subverted by the huge indebtedness accumulated during the 25 years of Reaganomics, which ended in September of 2008, and by accelerating fossil fuel unavailability, which signals the ironic end of petrochemical man.

Biocatastrophe Pie Chart

The following pie charts are our thumbnail sketches of a synergistically interrelated series of anthropogenic (manmade) excreta that are a result of and, in fact, characteristic of, late Holocene industrial man. The reader may reconfigure our thumbnail sketches at her/his leisure. As the Gulf oil spill disaster, an unexpected event, unfolds with ever increasing environmental, economic, social, and political impacts, new pie charts that incorporate this unfortunate event may be appropriate.

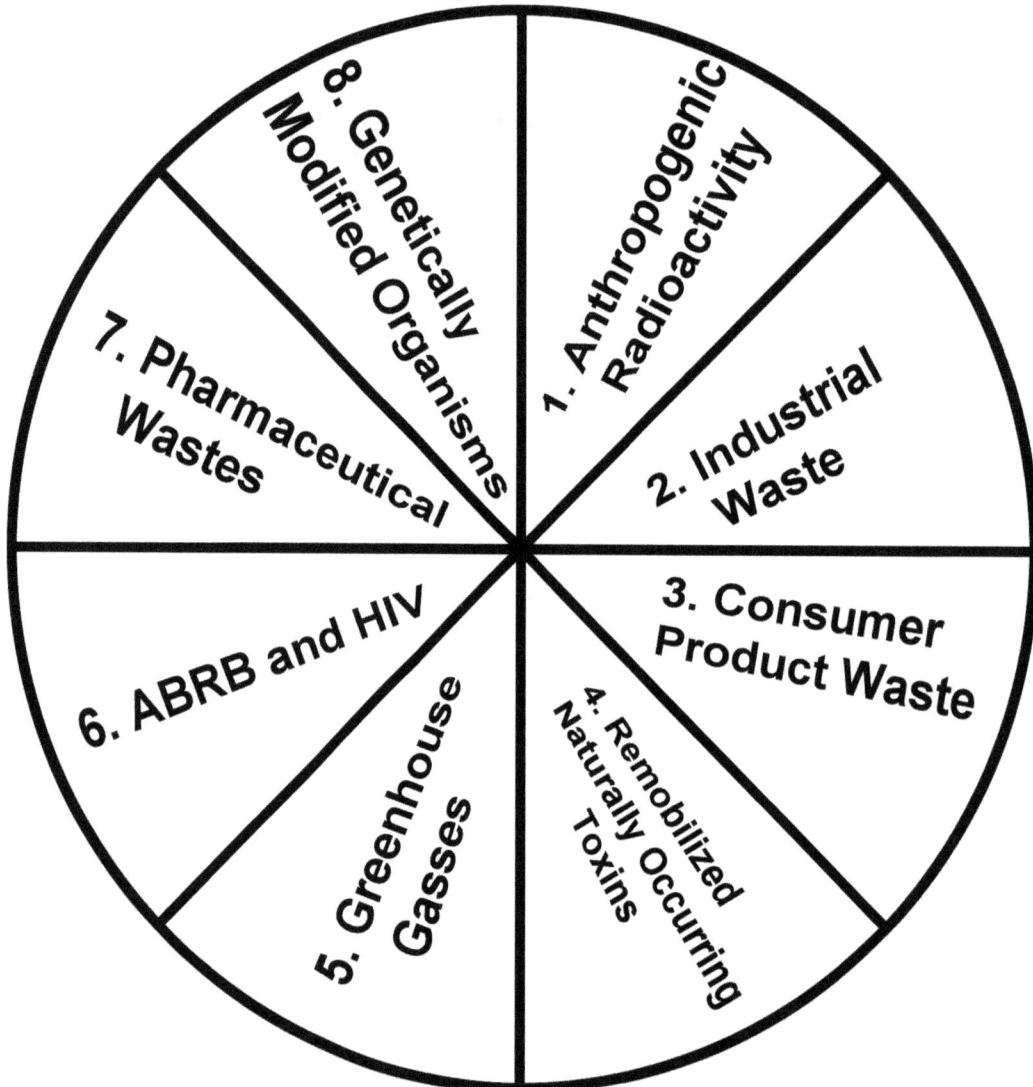

Figure 1

- ✓ **Anthropogenic radioactivity**
- ✓ **Industrial wastes**
- ✓ **Consumer product wastes**

- ✓ Remobilized naturally occurring toxins
- ✓ Greenhouse gases
- ✓ Antibiotic resistant bacteria (ABRB) and the HIV/AIDS virus
- ✓ Pharmaceutical wastes
- ✓ Genetically modified organisms

Chemical Fallout Pie Chart

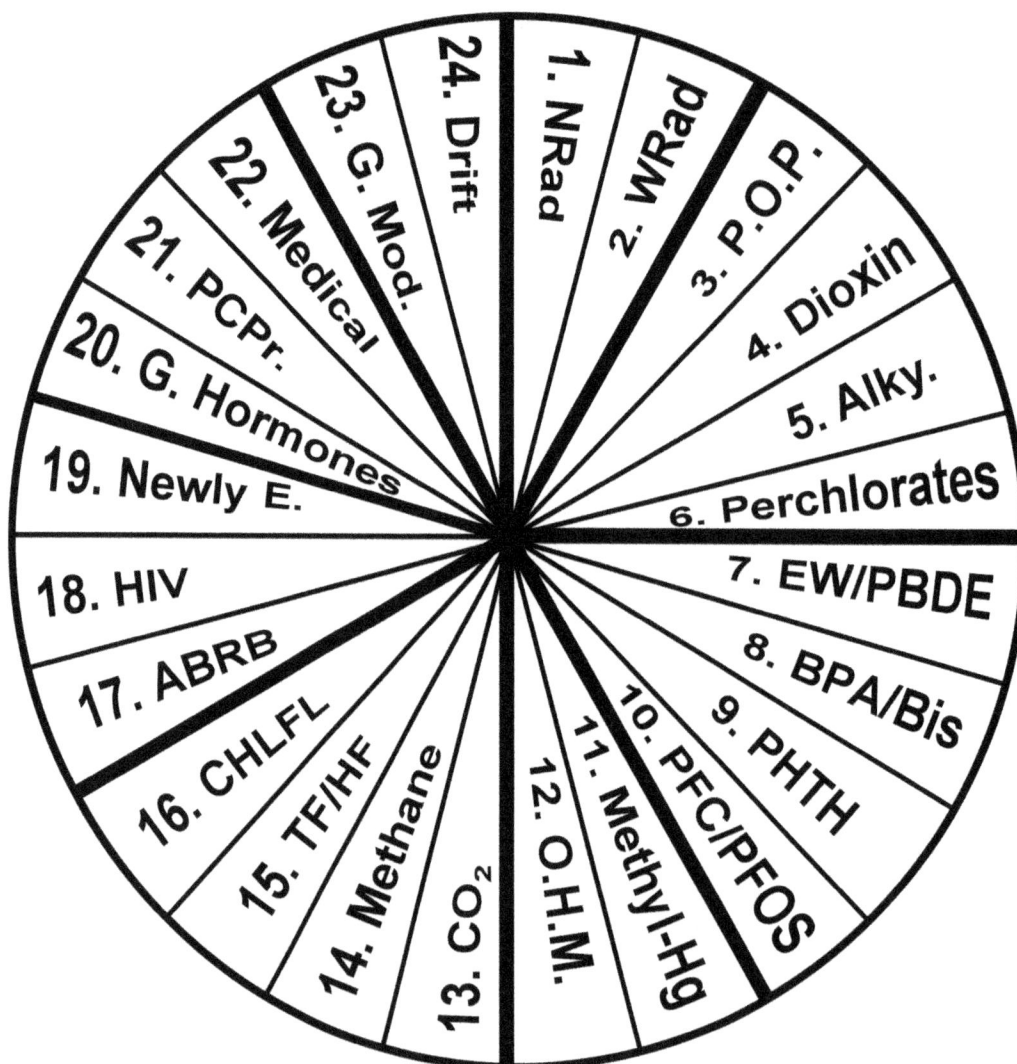

Figure 2

Key to the chart
Anthropogenic Radioactivity

1 **NRad**: Nuclear waste produced by the generation of nuclear electricity, including spent fuel and gaseous emissions

2 **WRad**: Nuclear waste produced by nuclear weapons testing and production and by the use of nuclear powered submarines and satellites

Industrial Waste

3 **P.O.P.**: Persistent organic pollutants derived from a multiplicity of industrial activities, including pesticides such as DDT, polychlorinated biphenyls (PCBs), and many other chemicals

4 **Dioxin**: Dioxin – furan cogeners, including those produced by the incineration of industrial chemicals, plastics, and a wide variety of consumer products, and now among the most widespread of all anthropogenic ecotoxins

5 **Alky.**: Alkylphenols, including nonylphenols (NPs), octylphenols (Ops), and nonylphenol ethoxylates (NPEs) used extensively as additives in plastics, emulsifiers, textile and carpet cleaning, as industrial detergents, and in pesticides

6 **Perchlorates**: Oxidizers used in rocket fuels and now omnipresent in municipal water supplies and as a cord blood ecotoxin

Consumer Product Waste

7 **EW/PBDE**: Brominated flame retardants used in the manufacture of computers, electrical appliances, vehicles, textiles, lighting and wiring, polystyrene insulation materials, and a wide variety of other consumer products.

8 **BPA/Bis**: Bisphenol-A (BPA), a ubiquitous component of polycarbonate and other plastics and a constituent of plastic bottles, mobile phones, medical devices, roofing panels, compact disks, epoxy resins, and many other consumer products

9 **PHTH**: Phthalates used as softeners in PVC products ranging from electrical cables to toys, floor coverings, vinyl wallpaper, rain clothing; also widely used as a solvent in personal care products, glues, printing inks, and paints

10 **PFC/PFOS**: Perfluorocarbon and Perfluorooctanesulfonic acid; industrial chemicals made from halogenated hydrocarbons, including those used in semiconductor photolithography, for protective coatings as surfactants in electronics, etching, medical uses, plastics, cooking utensils, cosmetics, and in firefighting foams

Remobilized Naturally-Occurring Toxins

11 **Methyl-Hg**: Methylmercury derived from the burning of coal, the use of medical and dental items, and other activities

12 **O.H.M.**: Other heavy metals, including arsenic, lead, cadmium, etc.

Greenhouse Gasses

13 **CO_2**: Carbon dioxide derived from the combustion of fossil fuels

14 **Methane**: Derived from permafrost melting, agricultural activities, and petroleum production

15 **TF/HF**: Tetrafluoromethane and hexafluoroethane, potent greenhouse gasses derived from silicon chip production

16 **CHLFL**: Chlorofluorocarbons, the release of which causes stratospheric ozone depletion

Pathogens

17 **ABRB**: Antibiotic resistant bacteria

18 **HIV**: Human immunodeficiency virus

19 **Newly E**: Newly emerging pathogens, such as SARS, Avian flu, H1N1, etc.

Pharmaceutical Wastes

20 **G Hormones**: Growth hormones utilized in food production

21 **PCPr**: Personal care product medications, such as Cialis, Prozac, and statins, which enter wastewater treatment systems and then enter the global atmospheric water cycle. Among the most biologically significant personal care products is triclosan, an antibacterial agent commonly used in soaps, detergents, deodorants, mouthwashes, toothpastes, cosmetics, food cutting boards, sportswear fibers, mattress pads, and many other consumer products

22 **Medical**: Medications, including antibiotics, which enter wastewater treatment systems and are then incorporated into the global atmospheric water cycle

Genetically Modified Organisms

23 **GMod**: Genetically modified organisms produced as herbicide or bacteria resistant plants

24 **Gdrift**: Genetic drift, the accidental release of genetically modified plants and animals

The Ecology of Money Pie Charts and Tables

The fundamental challenge of the advent of the age of biocatastrophe is to acknowledge, delineate, and analyze its components, and evaluate its social and economic costs. In a biosphere of dwindling natural resources, a contaminated global atmospheric water cycle, proliferating bacterial and viral infections, and growing populations of disenfranchised world citizens suffering food, water, and employment stress, the key question is, where are the financial resources that will be needed to mitigate the phenomenon of biocatastrophe? A sketch of national and world productivity and debt helps us understand the difficulty of meeting the financial challenges posed by the age of biocatastrophe.

The primary sources for the information pertaining to national and international gross productivity and debt are the *CIA World Fact Book*, International Monetary Fund, Federal Reserve Bank, Office of Management and Budget, and numerous other monetary information sources as cited in Wikipedia. This component of the text is an ongoing work-in-progress (WIP); additional information will be added as it becomes available. Updates, comments, and additional information as to the extent of National and World indebtedness are particularly welcome.

National Per Capita Debt in 2007

Country	Per Capita Debt in US Dollars
United States	$40,678.76
United Kingdom	$171,942.20
Germany	$54,477.50
France	$56,702.84
Italy	$40,328.31
China	$274.61
Japan	$11,708.07
Brazil	$1,207.30

(http://www.nationmaster.com/graph/eco_deb_ext_percap-economy-debt-extyernal-per-capita)

World and National Gross Domestic Productivity

	2007 list by the International Monetary Fund	2008 list by the CIA World Factbook	2009 list by the CIA World Factbook
World GDP	54.58 trillion	78.36 trillion	58.07 trillion
Domestic GDPs			
USA	13.84 trillion	14.33 trillion	14.43 trillion
Japan	4.38 trillion	4.84 trillion	5.11 trillion
China	3.28 trillion	4.22 trillion	4.81 trillion
Germany	3.32 trillion	3.81 trillion	3.27 trillion
Great Britain	2.80 trillion	2.78 trillion	2.22 trillion
France	2.59 trillion	2.98 trillion	2.67 trillion

	2007 list by the International Monetary Fund	2008 list by the CIA World Factbook	2009 list by the CIA World Factbook
Brazil	1.31 trillion	1.66 trillion	1.50 trillion
India	1.15 trillion	1.23 trillion	1.10 trillion
Indonesia	432.9 billion	496.8 billion	521 billion
Italy	2.10 trillion	2.39 trillion	2.11 trillion
Russia	1.28 trillion	1.75 trillion	1.23 trillion

A key element in the future viability of all national governments, including the American free enterprise system, is meeting the challenges of funding the legacy costs of medical care (in the US, Medicare and Medicaid) and Social Security. The European monetary crisis of 2010, which initially involved the issue of the solvency of Greece and soon also that of Spain, Portugal, Italy, and Great Britain, also centers on the immense social legacy costs of European governments with rapidly aging populations. Many European banks and governments were victimized by their participation in the toxic investment schemes and vehicles of the American-based shadow banking network. The challenge of funding the social legacy costs of both American and European governments, already a topic of acute concern, is greatly complicated by the wide variety of the current and future costs of the environmental and economic components of biocatastrophe. As the economic impact of an ever-growing population of the disenfranchised undermines a global economy characterized by limited and possibly declining growth and onerous debt obligations, what will be the sources of funding needed to meet the challenges of the age of biocatastrophe?

National Debt (ND) – A Sketch

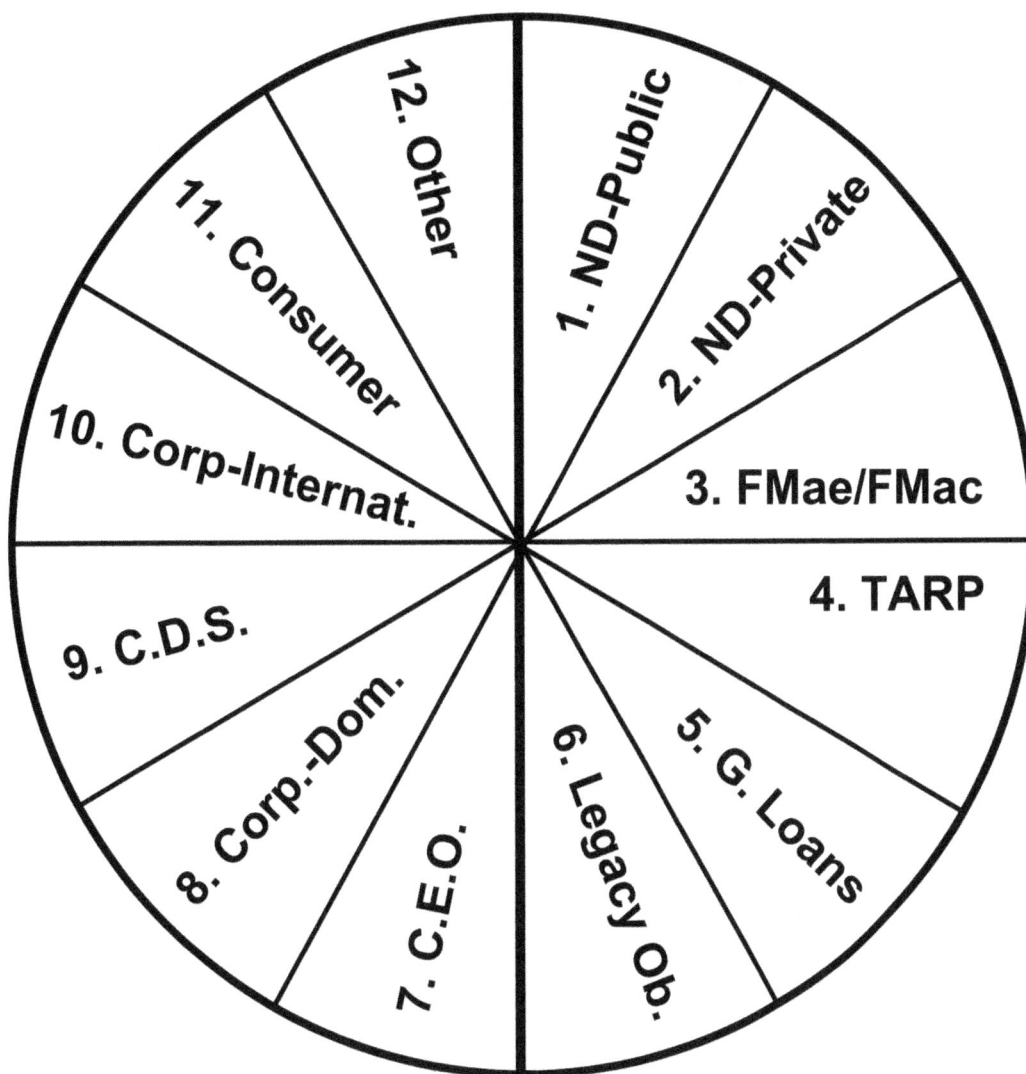

Figure 3

Public Debts	Rough estimates
1. **ND-Public**: US governmental debts, public*	7.5 trillion
2. **ND-Private**: US governmental debts, private*	5 trillion
3. **Fmae/Fmac**: Fannie Mae and Fannie Mac	2 trillion
4. **TARP**: Toxic asset relief program	1 trillion
5. **G. Loans**: Bond debt buyouts and guaranteed loans	2.5 trillion

Public Debts	Rough estimates
6. **Legacy Ob.:** Legacy obligations, including Social Security, Medicare, and Medicaid. A recent estimate of 100 trillion has been made (Ferguson 2009b).	8 trillion (40 trillion)
*Of the total US Government debt that is either privately or publicly held, 3.1 trillion is held by foreigners. This total does not include private corporate and banking debt: see 7 below.	
Total public debts	25.5 trillion
Private Debts**	
7. **C.E.O.:** External private debt owed to foreigners	10.6 trillion
8. **Corp.-Dom.:** Domestic corporate debt	?
9. **C.D.S.:** Credit default swap obligations	30 trillion?
10. **Corp-Internat.:** International debt owed by American corporations	?
11. **Consumer:** Consumer debts, including credit cards and personal loans	2.5 trillion
12: **Other:** Other consumer debts	?
Total private debts	+70 trillion

**In November of 2009, U. S. credit market debts, excluding national and personal debts, were estimated at 50 trillion dollars excluding credit default swap obligations. This component of indebtedness averages out to approximately $150,000 per U.S. citizen. The revised estimate by Niall Ferguson of legacy debt obligations of 100 trillion dollars for Social Security, Medicare, and Medicaid further illustrates the challenge facing the American economy of maintaining stability and viability in the 21st century.

The key unresolved question of the current financial crisis is the unknown amount of total indebtedness of American corporations, banks, hedge funds, bonds, credit default swaps, and other entities. Neither the federal government nor private analysts have yet to decipher the overlapping labyrinths of debt obligations held by private corporate interests, including the shadow banking network. For example, credit default swaps often overlap and cancel each other; 80 trillion in credit default swap transactions may actually result in only 35 trillion in credit default swap debts. A more detailed analysis of the many financial issues and controversies that have emerged since the demise of the Lehman Brothers in September 2008 can be accessed in the many books that have appeared since that date that are listed in the *History, Economics, and Politics Section* of the *Bibliographies* in *Volume 3*.

CAB: Credit Account Balance (National current account balances)

A national current account balance is the difference between any nation's export and import of goods and services, including government and private exchange.

Country	CAB in USD$, billions
China	371.833
Germany	252.501
Japan	210.967
Russia	76.163
Singapore	39.157
Sweden	38.797
Canada	12.726
France	-30.588
Italy	-52.725
United Kingdom	-105.224
United States	-731.214

(http://en.wikipedia.org/wiki/List_of_countries_by_current_account_balance)

United States Household Income Distribution

Bottom 10%	0 – 10,500
Bottom 25%	0 – 22,500
Top 25%	77,500 and up
Top 5%	167,000 and up
Top 1%	350,000 and up

(Source: U. S. Census Bureau statistics for the year 2005)

Epiphanies, News Bites, and other Data Relevant to National Debt

The following fragments of financial information and data help us fill in some of the blank spaces in a rational description of our current national financial crisis.

America's total national debt as of 2009 is 12.5 trillion dollars and is rapidly growing.

America's current account deficit in 2007 is 7% of gross domestic product.

Gross external debt in the United States is 13.77 trillion dollars (June 30, 2008).

United States public revenues in 2008 were 2.523 trillion dollars with expenses of 3.150 trillion dollars. United States GDP per capita in 2008: $46,800.

"At $140 trillion in 2005, the value of the world's financial assets hit a new peak and was more than three times as large as the total output of goods and services produced across the planet that year."
(bigpicture.typepad.com/comments/2007/01/worlds_assets_h.html).

"In the United States, wealth is highly concentrated in a relatively few hands. As of 2001, the top 1% of households (the upper class) owned 33.4% of all privately held wealth, and the next 19% (the managerial, professional, and small business stratum) had 51%, which means that just 20% of the people owned a remarkable 84%, leaving only 16% of the wealth for the bottom 80% (wage and salary workers)."
(http://sociology.ucsc.edu/whorulesamerica/power/wealth.html).

"In the United States at the end of 2001, 10% of the population owned 71% of the wealth, and the top 1% controlled 38%. On the other hand, the bottom 40% owned less than 1% of the nations wealth." (http://en.wikipedia.org/wiki/Distribution_of_wealth).

In the United States "roughly 2.245 million households, or the top 1.9 percent, had incomes greater than $250,000 in 2007." (Tomasky 2009, 18).

"The ratio of CEO pay to factory worker pay rose from 42:1 in 1960 to as high as 531:1 in 2000, at the height of the stock market bubble, when CEOs were cashing in big stock options. It was at 411:1 in 2005. By way of comparison, the same ratio is about 25:1 in Europe." (http://sociology.ucsc.edu/whorulesamerica/power/wealth.html).

"Government debt can be categorized as internal debt, owed to lenders within the country, and external debt, owed to foreign lenders."
(http://en.wikipedia.org/wiki/Public_debt).

"Worldwide debt levels are perhaps worth two or three years of GDP. GDP (at currency exchange rate) was $40 trillion during 2004. Debt levels may therefore be about $100 trillion." (http://en.wikipedia.org/wiki/Global_debt).

"The two GSEs [Fannie Mae and Freddie Mac] have approximately US$ 1.5 trillion in bonds outstanding."
(http://en.wikipedia.org/wiki/Federal_takeover_of_Fannie_Mae_and_Freddie_Mac).

"The entire CDS [Credit default swap] market has a notional value in the vicinity of US$ 62 trillion."
(http://en.wikipedia.org/wiki/Federal_takeover_of_Fannie_Mae_and_Freddie_Mac).

In 2007, the US credit market debt was 350% of the gross domestic product (GDP) (JW).

144

World of Plastics Pie Charts

Figure 4

The following two pie charts were retrieved by the Liberty Tool Company from the estate of a Framingham, MA, engineer with a specialty in plastics, plasticizers, and the manufacture of Air Force pilot helmets and other safety apparatus. These two pie charts provide an interesting insight into the world of plastics and provide the names of many of the principal 1950s creations of a highly inventive petrochemical industry, the ecotoxic legacy of which is now an important constituent of the atmospheric water cycle, including maternal cord blood. The outer circle of Figure 4 provides basic delineations of many of the principal products of the petrochemical industry. It is interesting to note some of the chemicals in the secondary circle of names: the notorious Teflon as a "special" plastic, the equally well known Bakelite as a polyethylene derivative. Noted on the outer circle of this pie are: phenolic, urea, melamine, ethyl cellulose, cellulose acetate, cellulose acetate butyrate, nylon, special, resin acrylonitrile, polyethylene, vinyl chloride acetate, vinyl chloride, methacrylate, and styrene.

The reverse side of this particular pie-guide provides information on the thermal plastic qualities of these petroleum products, including compression and injection data, specific gravity, heat distortion, price per pound (e.g. vinyl chloride had a price range between $.40 and $.90 a pound), and coloring possibilities.

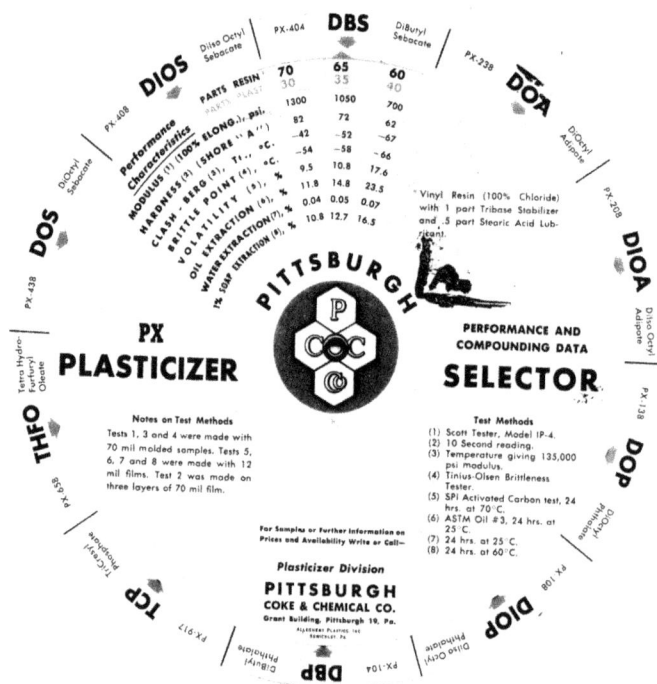

Figure 5

DBS — DiButyl Sebacate — PX-404
DIOS — Diiso Octyl Sebacate — PX-408
DOA — DiOctyl Adipate — PX-338
DOS — DiOctyl Sebacate — PX-438
DIOA — Diiso Octyl Adipate — PX-208
THFO — Tetra Hydro-Furfuryl Oleate — PX-638
DOP — DiOctyl Phthalate — PX-138
TCP — TriCresyl phosphate — PX-917
DIOP — Diiso Octyl Phthalate — PX-108
DBP — DiButyl Phthalate — PX-104

PX PLASTICIZER SELECTOR

PITTSBURGH COKE & CHEMICAL CO.

PERFORMANCE AND COMPOUNDING DATA

Performance Characteristics	DIOS	DBS	DOA
PARTS RESIN / PARTS PLAST.	70 / 30	65 / 35	60 / 40
MODULUS (1) (100% ELONG.), psi.	1300	1050	700
HARDNESS (2) (SHORE "A")	82	72	62
CLASH-BERG (3), Tf, °C.	-42	-52	-67
BRITTLE POINT (4), °C.	-54	-58	-66
VOLATILITY (5), %	9.5	10.8	17.6
OIL EXTRACTION (7), %	11.8	14.8	23.5
WATER EXTRACTION (6), %	0.04	0.05	0.07
1% SOAP EXTRACTION (8), %	10.8	12.7	16.5

Vinyl Resin (100% Chloride) with 1 part Tribase Stabilizer and .5 part Stearic Acid Lubricant.

Notes on Test Methods

Tests 1, 3 and 4 were made with 70 mil molded samples. Tests 5, 6, 7 and 8 were made with 12 mil films. Test 2 was made on three layers of 70 mil film.

Test Methods

(1) Scott Tester, Model IP-4.
(2) 10 Second reading.
(3) Temperature giving 135,000 psi modulus.
(4) Tinius-Olsen Brittleness Tester.
(5) SPI Activated Carbon test, 24 hrs. at 70°C.
(6) ASTM Oil #3, 24 hrs. at 25°C.
(7) 24 hrs. at 25°C.
(8) 24 hrs. at 60°C.

For Samples or Further Information on Prices and Availability Write or Call—

Plasticizer Division
PITTSBURGH COKE & CHEMICAL CO.
Grant Building, Pittsburgh 19, Pa.

The second pie-guide (Figure 5, dated 1952) provides data pertaining to the performance characteristics of plasticizers produced by the Pittsburgh Coke & Chemical Co. Of particular concern are modulus (percentage of elongation), hardness, clash-berg as expressed by temperature, brittle point also expressed by temperature, volatility, oil extraction, water extraction, and 1% soap extraction. All these performance characteristics and chemical and physical properties are a commentary on the usefulness of the many plastics that played such an important role in the growth and expansion of global military/industrial/consumer society (GMICS). All of these plastic compounds are a product of the petrochemical industry, the chemistry of which is briefly discussed at the beginning of the next *Volume*. While the industrial agricultural chemicals DDT, chlordane, heptachlor, toxaphene, hexachlorobenzene, aldrin, dieldrin, endrin, and mirex have long been among the most well known ecotoxins produced by the petrochemical industry, the pie charts illustrated above are a commentary on the emerging issue of ecotoxicity of personal care products and other consumer goods, many often made of or containing plastics or plasticizers. While the above pie-guides and the chemicals they delineate may date as early as the late 1950s, the issue of the many toxic byproducts derived from the use and disposal of personal care products, fabrics, clothing and the incineration of plastics has only recently emerged. The publication of Theo Colburn's *Our Stolen Future* (1996), followed by Sheldon Krimsky's *Hormonal Chaos* (2000) were the environmental equivalents of Rachel Carson's *Silent Spring* (1962). These texts were the first comprehensive analyses of the role many plastics-derived consumer product ecotoxins play in the etiology of the health physics impact of endocrine disrupting chemicals (EDCs). Many of these chemicals, such as phthalates and bisphenol-A, are constituents of consumer products commonly found in everyday domestic and work environments, clothing, transportation vehicles, and the shopping malls of global consumer society. A wide diversity of other ecotoxins are derived from the incineration of these plastics and consumer products. Only a tiny percentage of modern day plastics and consumer products are properly recycled, with a minimal external cost to their production and use. The rapid growth of a large number of megacities in developing nations, some of whose residents may unfortunately mimic the lifestyles of America's object and wealth-craving consumers, ensures that huge quantities of what are now considered to be socially and economically useful chemicals, plastics, brominated fire retardants (PBDEs), and personal care products will continue to be the source of massive amounts of non source point pollution, a now increasingly visible tip of Hotel California's round-world iceberg of bioactive pollutants. Dioxin furan congeners are among thousands of other invisible or nearly invisible products of a modern military industrial consumer society that has surpassed the limits of ecological sustainability. The thousands of chemicals whose formulas are proprietary information join with the accelerating proliferation of nanotoxins to form an invisible counterpart to the spectacle of the Gulf oil spill disaster as a regional biocatastrophe. The third *Volume* in this series provides an introduction to the chemistry, toxicity, and biomonitoring of these and other anthropogenic ecotoxins, the legacy of our late 19th century commitment to petrochemistry.

www.ingramcontent.com/pod-product-compliance
Lightning Source LLC
Chambersburg PA
CBHW051218200326
41519CB00025B/7168